DIGITAL RESEARCH METHODS AND THE DIASPORA

The computational turn in the social sciences and humanities has generated much excitement about the potential to refresh our approaches to the study of the techno-social. From natively digital to digitised data, researchers of digital diasporas increasingly find themselves working with a range of disparate digital objects. These digital objects can include anything from hyperlink to timestamps, from platform behavioural metrics such as react, share, or retweet to different media formats such as text, image, pre-recorded or livestreamed videos.

Taking these disparate objects into account, this book introduces digital methods as research strategies not only for dealing with the ephemeral and unstable nature of tracing the diaspora with digital data, but also for reconceptualizing digital diasporas as assemblages and networks of more-than-human actors. The book also introduces a range of theoretical perspectives and methodological techniques to studying digital diasporas as contingent and processual hybrid collectives of heterogeneous material, cultural, and practice-based assemblages.

This book will be essential reading for students and scholars interested in the digital space and transnational communities.

Dang Nguyen is a Research Fellow at the ADM+S Centre at RMIT University, Australia. She researches the social implications of digital technologies by bringing together methods of different disciplines and looking beyond Western contexts. Dang has published in venues such as *New Media & Society*, *Information, Communication & Society*, *Social Science & Medicine*, and *Health & Place*.

Digital Diaspora Series

Series Editors: Anh Nguyen Austen[1] and Anita Lucchesi[2]
[1]*Research Associate, Centre for Refugees, Institute for Humanities and Social Sciences, Australian Catholic University;* [2]*Postdoctoral Research Associate, Luxembourg Centre for Contemporary and Digital History*

This series explores the nexus of digital life, migration, and diaspora studies in the humanities and social sciences. The books in the series include methods, theories, and case studies about how migrants and diasporas engage in technology to deal with conflicts, displacement, climate change, and everyday life. Topics include the complex roots, forms, and implications over time of how digital communities, diasporic connections, and migrant cultures inform and produce new knowledge across disciplines including data and network science, media and communications, ethnography, anthropology, geography, psychology, history, literature, political science, and education. Interdisciplinary methods and perspectives from the global south and transnational studies are of particular interest.

Digital Research Methods and the Diaspora
Assembling Transnational Networks with and beyond Digital Data
Dang Nguyen

DIGITAL RESEARCH METHODS AND THE DIASPORA

Assembling Transnational Networks with and beyond Digital Data

Dang Nguyen

LONDON AND NEW YORK

Designed cover image: Cover Image designed by DALL-E and Tam Nguyen

First published 2024
by Routledge
4 Park Square, Milton Park, Abingdon, Oxon OX14 4RN

and by Routledge
605 Third Avenue, New York, NY 10158

Routledge is an imprint of the Taylor & Francis Group, an informa business

British Library Cataloguing-in-Publication Data
A catalogue record for this book is available from the British Library

ISBN: 978-1-032-37349-2 (hbk)
ISBN: 978-1-032-37348-5 (pbk)
ISBN: 978-1-003-33655-6 (ebk)

DOI: 10.4324/9781003336556

Typeset in Sabon
by MPS Limited, Dehradun

For Mum and Dad.
Con thương Ba Mẹ nhiều lắm.

CONTENTS

FIGURES

TABLES

ACKNOWLEDGEMENT

I wish to thank Julian Thomas and the TCP Lab team for convincing me that I can, and should, write a book. Anh Nguyen Austen suggested the idea of a methods book for digital diaspora studies, which prompted me to pursue this project. Michael Arnold encouraged me to be experimental with my research methods both here and elsewhere, and convinced me that writing a methods book is in fact a very good idea. Tam Nguyen made sure that all the figures and tables in this book are made to the best standard, and is the most patient and talented graphic design expert a writer could ask for. Thi Nguyen remains the best copyeditor; all mistakes are my own.

INTRODUCTION

The computational turn in the social sciences and humanities has generated much excitement about the potential to refresh our approaches to the study of the techno-social. From natively digital to digitised data, researchers of digital diasporas increasingly find themselves working with a range of disparate digital objects. These digital objects can include anything from hyperlinks to timestamps, from platform behavioural metrics such as react, share, or retweet, to different media formats such as text, image, and pre-recorded or livestreamed videos. Digital data often become of interest to researchers as heterogeneous assemblages. These assemblages can be mobilised to tell insightful stories about the changing sociality of diasporas. They also often lead researchers to unfamiliar territories, requiring them to consider the full agency of technology in mediatising transnational social relations. This book not only presents digital methods as research strategies to cope with the evanescent and unstable nature of tracing the diaspora through digital data, but it also reimagines digital diasporas as assemblages and networks of more-than-human actors. Additionally, the book introduces various theoretical perspectives and methodological techniques for studying digital diasporas as contingent and processual assemblages, rather than static communities that merely fluctuate over time.

Digital research methods are often experimental and situational, accommodating for the ever-changing landscape of the platforms on and through which diasporas are enacted. Researchers from across the social sciences and humanities are showing increasing interest in utilising innovative methods to analyse both natively digital and digitised data, in order to gain a more sophisticated understanding of the intersection of technology and society. Given their transnational nature and increasingly digital characteristics, diasporas are well-suited to benefit from these methodological innovations. The 21st-century diaspora is in a constant state of adaptation and evolution, shaped by the technologies that enable and constrain its formation, expansion, and contraction. Digital diasporas can be understood as networks of ideas, people, materials, and money that are more than human; a commitment to assembling the heterogeneous traces left by these networks can help foster new innovative research programs, build bridges between disciplines and communities, and create a more vibrant and inclusive field.

DOI: 10.4324/9781003336556-1

My motivation for writing this book comes from my own multidisciplinary training in media studies, internet studies, and the history and philosophy of science and technology. Throughout my research training as a Vietnamese diasporic researcher who physically travelled across the English-speaking world – and metaphorically across the disciplines that were gracious enough to host me, including those that did so in a less formal capacity such as anthropology, sociology, cultural studies, and Vietnamese studies – I have found research methods to be the most productive entry point into facilitating multi- and interdisciplinary conversations. While it might be the case that different disciplines are mortgaged to different theoretical perspectives and methodological approaches, it is also the case that many researchers are genuinely interested in learning about what researchers trained in other disciplines 'do' when they do research. Digital research methods often generate as much excitement and curiosity as they do trepidation and scepticism; while graduate and emerging researchers are often drawn to the idea of experimenting with automated methods so that they can mobilise large amounts of data towards telling interesting stories, they also often tend to be discouraged by the various barriers to acquiring new technical skills and/or reconciling with unfamiliar theoretical sensibilities. Many researchers, myself included, actively work to engage with methods from a wide range of disciplines, both to make ourselves competitive in the job market and as a means to say something new about the topics we study. Innovating through methods is an increasingly popular strategy for early-career researchers hoping to establish themselves in the various academic disciplines they are contributing to, and digital research methods make up an exciting repertoire with which to do this.

The purpose of this book is not to provide tutorials on using digital tools that may soon become obsolete due to the rapidly evolving nature of platforms and their algorithmic recommendation systems. Instead, it introduces theoretical perspectives and methodological techniques for studying digital diasporas as hybrid collectives of heterogeneous material, cultural, and practice-based assemblages. By arguing for a hybrid approach to assembling our rich techno-social worlds as digital diaspora assemblages, this book not only refreshes the way we view diasporas but also seeks to chart some promising directions for doing digital diaspora research in methodologically innovative ways. It does so primarily through weaving examples of mixed-method digital research that either I have conducted myself, or exemplars of noteworthy research projects related to digital diaspora, at the end of each chapter (except for Chapter 6). These examples are accompanied by discussions of methodological and theoretical issues related to their approaches – written at the introductory level for graduate students and researchers without any prior knowledge of digital research. As such, this book aims to both be useful to researchers seeking to learn how to do digital research in relation to diaspora studies and to act as an intervention in the way we think about digital diasporas.

The book is organised into two parts. **Part I** (chapters 1–3) presents quantitative approaches to studying digital diasporas by introducing network theories and network analysis – as well as discussing linguistic issues in studying digital diasporas by introducing natural language processing principles and applications. Chapter 1 gives an overview of the historical roots of network analysis and its multidisciplinary practice in contemporary scholarship before discussing the tensions between the relational and the formalist approaches to network theory. Arguing that both approaches can be operationalised to assemble digital diasporas using a variety of data types, the chapter proceeds to discuss the pitfalls of the quantification of qualitative phenomena – such as diaspora – and the kind of data required for such a task. The chapter ends with a discussion of the e-Diasporas Atlas project as a case study of the quantification of diasporic networks using web data.

Chapter 2 introduces social network analysis as a method to assemble and examine digital diasporas by discussing a range of basic network statistics and theories that are useful to researchers of diaspora. The chapter also presents a case study of how these statistics and theories can be applied to the study of Vietnamese traditional medicine networks on Facebook before concluding with a discussion on the opportunities and challenges that social network analysis poses for the study of digital diaspora, highlighting the deliberative nature of this method in setting parameters around constructed networks and outlining the theoretical implications that this method could have for diaspora studies.

Chapter 3 introduces a range of natural language processing techniques and models and discusses the strengths and weaknesses of NLP techniques and models in the analysis and synthesis of large-scale social media datasets – with a particular focus on what they may mean for researchers of the digital diaspora. The chapter also argues that more attention needs to be paid, and investment made, to the development of NLP methods and libraries for non-English languages to aid researchers with the task of studying large multilingual corpora generated as digital traces left behind by diasporic communities on social media. By analysing a Vietnamese social media dataset as a case study, this chapter examines how NLP is a method sensitive to detecting diasporic activities. The chapter also explores how NLP can help researchers reconcile with the relational paradigm in network analysis by focusing on the content of diasporic exchanges and concludes by outlining the challenges and drawbacks of implementing NLP on non-English datasets.

Part II (chapters 4–6) presents qualitative and technological approaches to studying digital diasporas by introducing digital ethnography and discussing emergent methodological issues for digital diaspora research in the presence of algorithmic recommendation systems. Chapter 4 interrogates digital diasporas not as naturally occurring research sites, but as complex arrangements of difference that require a new vocabulary. The chapter suggests that researchers must approach digital diasporas as ongoing enactments of shared identity, common meaning, and bounded sovereignty subject to transcension. Following this, researchers should also be prepared to attend to disruptive events and unfamiliar territories, especially considering technology's non-human agency in mediating transnational social relations.

Chapter 5 presents digital ethnography as a methodology for studying diasporas, allowing researchers to engage with human-machine landscapes and include non-human actors in examining cultural formations and information flows. Digital ethnography acknowledges the digital as a theoretical and practical aspect of our world and can be applied across research questions. The chapter also discusses how digital ethnography's adaptive nature accommodates the complexities of the diasporic situation through hybrid ethnography and complementary methods. Digital ethnography's non-conformity principle allows researchers studying digital diasporas to engage with technology creatively and innovatively: this includes not only using digital technologies during fieldwork, like video ethnography, but also by mobilising, deconstructing, and theorising digital trace data in ways that allow us to fundamentally rethink technological processes.

Chapter 6 explores how automated systems act as infrastructure for the emergence of digital diasporas and considers the methodological implications for researchers. Recommender systems structure how people discover information and communities online by matching users with digital resources based on past behaviour and user preferences. Despite differing goals, recommender systems serve as cultural intermediaries that organise user experiences with digital services. The chapter provides an overview of recommender systems and discusses their impact on digital diaspora research, specifically in relation to social recommender systems and recent advances in deep learning.

This book was written with the dual goal of bringing together seemingly incommensurable methodological approaches under the same overarching research program and introducing these approaches to researchers of diaspora studies. Beyond this goal, the book was also written to convey a sense of urgency around thinking seriously *with* and *through*, rather than *about* technologies in relation to diasporas – as if they can be abstracted away from diasporic practices. The 'digital' in digital diasporas has always been simultaneously technical and social; it is through investigating the various intricate layers of automation that have been built into the architecture of digital diasporas that researchers can complicate and problematise prevailing narratives around the impact of technology on diasporic lives. Digital diaspora research carries immense potential for innovation; this book is an attempt at sketching what that potential might look like if we begin to integrate technical accounts into our usual social accounts of the digital – and commit to never separating the two.

PART I

Quantitative approaches: making natively digital data speak

1

DIASPORAS AS QUANTIFIABLE NETWORKS

Network analysis as a multidisciplinary inquiry

Network analysis as a 21st century scholarly practice is multidisciplinary in nature. Even though the study of networks has a long history in mathematics and natural sciences (cf. Eulerian path and the problem of Seven Bridges of Königsberg[1]; Cayley's labelled tree and graph enumeration[2]), contemporary scholarly practice of network analysis spans engineering, computer science, linguistics, sociology, history, political science, media studies, internet studies, and other emerging disciplines. Naming network science 'the science of the 21st century' and attributing the emergence of network science to the increased availability, accessibility, and capacity of computational tools and data, Barabási (2016) traced the explosive interest in networks during the first decade of the 21st century by plotting the number of citations of two classic papers – Erdős-Rényi (1959)[3] and Granovetter (1973).[4] After decades of only limited impact outside their fields despite being highly regarded within their respective disciplines, these two classics have gathered significant interdisciplinary attention since the early 2000s, marking the popularisation of the ideas they discussed – that of universal properties in random graphs, and that of the strength of weak ties in social networks.

Since the turn of the century, research drawing on network techniques and concepts has proliferated across empirical domains and interests. The concept of network contagion, for example, was the subject of a hugely controversial paper on massive-scale emotional contagion through social networks by Kramer et al. (2014), in which the authors conducted a massive (N = 689,003) experiment on Facebook to demonstrate that people's emotional states can be transferred to others via emotional contagion – wherein people experience the same emotions without their own awareness. The study was widely criticised for its ethical implications and inspired debates into the responsibilities of academic researchers to protect the vulnerability of people who disclose personal information on social media, and who are unable to control how their personal information is presented in technologically mediated environments. Two years earlier, the concept of social contagion was also the subject of another paper that utilised

DOI: 10.4324/9781003336556-3

Facebook as a network experiment environment, where researchers from the University of California, San Diego conducted a randomised controlled trial of political mobilisation messages delivered to 61 million Facebook users during the 2010 US congressional elections (Bond et al., 2012). These researchers found that their targeted messages not only directly influenced political self-expression, information seeking, and real-world voting behaviour of the millions of users who received them, but also indirectly influenced their friends, and their friends of friends. The paper also highlighted how the effect of social transmission on real-world voting was greater than the direct effect of the messages themselves and explained this by invoking another network concept – that of the strong tie between close friends as instrumental for spreading both online and real-world behaviour in human social networks. The authors estimated that their 'get out the vote' message yielded 300,000+ more voters at the polls in 2010. They replicated this experiment during the US Presidential Election in 2012 and once again observed a significant increase in voting, including among the close friends of those who received the message to go to the polls (Jones et al., 2017).

Network analysis is not just about conducting experiments on social media, however. In her highly original book, *Between Monopoly and Free Trade: The English East India Company, 1600–1757*, Emily Erikson (2014) conducted a systematic quantitative network study based on primary sources into how private traders channelled the growth and success of the East India Company (EIC) – one of the most powerful and enduring organisations in history. By mobilising network concepts such as embeddedness and structural cohesion, Erikson argued that it is the decentralised organisational structure of the EIC as constructed through the combination of private and company trade that explained its continued expansion and adaptability over nearly two centuries as a predominantly commercial operation. The success of the EIC, as such, is attributable not to imperialism or the centralisation of administrative forms, but rather to the systematic effects on the conduct of the EIC trade through decentralised, lateral information transfer. This decentralisation is made possible through a tighter network of communications that put significant autonomy into the hands of local agents of the EIC. Network as a conceptual approach and as a method, therefore, offers original and rigorous approaches with which researchers can measure and investigate specific social mechanisms to reveal how individual lives intersect and culminate into larger social and organizational structures – as well as historical patterns. Quite different from the natively digital, 'found' data of social media, historical records of ships' journals and logs, ledgers, receipts, absence books, pay books, and correspondence can be curated, digitised, and made computable so that they can be mobilised for network analysis.

Network also offers an insightful approach with which to assess partnership effectiveness in community development. Pope and Lewis (2008) used a network approach to examine what partnerships require in terms of governance if they are to be effective, by focusing on network structures and relationship building among ten different public service partnerships in Victoria, Australia. To collect data, they conducted 120 interviews of up to 15 members of the partnership's steering committee or equivalent governance body, either face-to-face or via telephone. Interviewees were asked a range of questions that would enable the researchers to construct network maps of the partnerships. The questions include their relationships with each of the other partners, organisations or people that should have been involved that were not, the biggest successes or achievements that had resulted from the partnership, what had helped and what had hindered the partnership's work, what could have been done better, what lessons they had learned that could be translated elsewhere, and sustainability of the partnership activities or its outcomes. Most importantly, interviewees were also asked to name the people they talked to most to undertake their day-to-day work in the partnership and to get strategic information about the partnership – a technique known as name generation – to create network maps. Network concepts such as centrality and brokerage[5] were then mobilised to evaluate the governance aspects of these partnerships.

It can be seen from the above sample of research that network as a theoretical and methodological approach is both generative in the ideas it inspires and accommodating in the research purposes it serves. While different disciplines naturally have diverging agendas, priorities, and interests, many of the problems they face share commonalities that lend themselves to the cross-disciplinary fertilisation of tools and ideas. As a result, researchers working on problems specific to their disciplines may find utility and inspiration in network as a research paradigm and as a conceptual anchor. There are some very intuitive ways in which network analysis could enrich diaspora studies, given the themes of dispersal, community, (dis)connectedness, culture, and difference in diaspora research. With global diasporas becoming increasingly networked – in the sense that diasporic actors are increasingly engaging with digital technologies in the enactment of everyday diasporic lives – thinking about diasporas through a network lens not only offers novel theoretical and analytical insights into the phenomenon of diaspora but also allows researchers to properly account for the technology question in 21st century diasporic lives, which has thus far been inadequately addressed. The case study at the end of this chapter will demonstrate how the emergence of e-diasporas can be studied in tandem with the diffusion of the Internet and the development of multiple online public services as networks. For now, I will turn to a discussion of Actor-Network Theory and its capacious theoretical sensibilities as a fruitful common point of departure for diaspora and network research.

Actor-network theory and the sociology of Gabriel Tarde

Despite its name, Actor-Network Theory (ANT) is not so much a theory of network, but rather a theoretical movement[6] that instructs particular types of observation about how humans and non-humans assemble, come apart, and then come back together again – processes that resemble the formation and dissolution of networks. ANT views sociality not only on the level of association between humans and non-humans, but also from a dynamic of interaction between entities and categories, all of which produces problematic effects (Latour, 2005). ANT asks us to rethink the 'social' as bundles of ties that can be mobilised to account for some other phenomenon; it is only through following the 'actors' who interact with each other while leaving behind traces of network among themselves that the 'social' can properly be assembled. In other words, from a Latourian view, the social and network are one and the same: the social is never fixed, and there does not exist a social world outside of networks of relations among actors. This is what is meant by actor-networks: they are assemblages whose existence ceases as soon as the actors involved stop performing the interactions that characterise their relations.

For ANT, the autonomy and the relations of any entity are understood as an accomplishment. This accomplishment is likely to depend on supporting arrangements and support acts, which are not just 'social' but include environmental entities previously categorised as nature (Latour, 2005). Central to ANT is the insight that actor-networks are accomplishments that must be explained rather than assumed (Baiocchi et al., 2013). Diasporic networks, from an ANT point of view, cannot be assumed to exist *ex ante*; it is only through following diasporic actors and carefully tracing and mapping their relations that a diasporic network can be said to exist. While the instruction to 'follow the actor' in ANT neither explains nor outlines how this work is to be done, it suggests that the empirical field of work continuously emerges as a result of researcher engagement. In this way, ANT encourages explorative and experimental ways of attuning to the world in that it underlines how particular modes of relating, though counter-intuitive to the existing ways in which we have come to understand human sociality, can open up the interconnected, consubstantial, co-constituted ways in which humans and non-humans mutually shape each other in and through their encounters (Lury, 2021; Mol, 2008). A thorough and committed critique of social theory, ANT deploys a range of shifting vocabulary and mantras to guide ANT researchers through the task of assembling the social: quasi-actants, hybrids, immutable mobiles, matters of fact vs. matters of concern, intermediaries vs. mediators, translation, trail, alliance, sociology of translation vs. sociology of association vs. sociology of the social (Latour, 2005). This array of concepts destabilises a core framework with which researchers could readily rely on, so that theories coming out of ANT are always 'alive'. As John Law (1999) – another major proponent of ANT alongside Bruno Latour – puts it:

> For these attempts to convert actor-network theory into a fixed point, a specific series of claims, of rules, a creed, or a territory with fixed attributes also strain to turn it into a single location. Into a strongpoint, a fortress, which has achieved the double satisfactions of clarity and self-identity. But all of this is a nonsense for, to the extent that it is actually alive, to the extent that it does work, to the extent to which it is inserted in intellectual practice, this thing we call actor-network theory also transforms itself. This means that there is no credo. Only dead theories and dead practices celebrate their self-identity. Only dead theories and dead practices hang on to their names, insist upon their perfect reproduction. Only dead theories and dead practices seek to reflect, in every detail, the practices which came before. (p. 10)

This network dynamism goes hand in hand with ANT's observation that action is distributed and redistributed between humans, artefacts, and the environment – that non-human actors have their own agency and independent activity. When an actor commits an action, they are not the sole source of such an act: action is never taken out of full agency, and the nature of agency needs to be disentangled by the same process that appoints actors. It is in this sense that action should always be studied relationally: an actor is never alone in acting. As Latour (2005) puts it:

> Action is borrowed, distributed, suggested, influenced, dominated, betrayed, translated. If an actor is said to be an actor-network, it is first of all to underline that it represents the major source of uncertainty about the origin of action. (p. 46)

This does not imply, however, that actors do not have agency. An ANT account of sociality instead relies on an explicit theory of action that demands a kind of transparency in agency, wherein agency needs to first be identified as 'doing' and making a visible difference – as changing a state of affairs. Consequently, ANT disallows 'invisible agency'; accounts of sociality that entail an unaccountable or invisible social force that mysteriously drives action are considered invalid in ANT terms. ANT favours a 'concatenation of mediators' (Latour, 2005, p. 59) as a means to draw social cartographies: each and every single node of a network should be said to fully act in relation to each other. Latour (2005) goes as far as saying that due to this dynamism, perhaps ANT networks could even be rebranded as 'worknets' or 'action nets', so that the labour that goes on in laying down net*works* can be emphasised – and so that the distinction between the technical network as an active mediator and the ANT network as a stabilised set of intermediaries can be made as apparent as possible.

While network as elaborated in ANT has unique characteristics that are not found in graph theory or network topology, its embrace of the quantification of social phenomena shares the mathematical lineage of graph theory in that ANT approaches its subjects from up close, rather than from a distance. Graphs are mathematical structures used to model pairwise relations between objects: Euler (1736) solved the problem of the seven bridges of Königsberg by laying out their positions in relation to each other through symbolic representations, while Cayley (1857) provided a visual overview of the logical branching that occurs when iterating the fundamental procedure of differentiation by introducing the concept of the mathematical 'tree'. A graph's elementary component is the monad – an individual unit with built-in relational qualities – rather than an atom, whose multiplication is always oriented towards the outline of emerging structures as aggregates and as wholes. In a similar sense, actor-networks set out to bypass the notion of structure as an aggregate by following the actors – whose traces are rendered visible and present by means of their interactions with each other – towards composite assemblages as provisional accomplishments. The distinction between network as assemblage and structure as aggregate is, in Latour (2010)'s observation, an artefact of where the observer is placed and the number of entities they are considering at once. The gap between the overall structure and the underlying components, argues Latour, is due to a deficit of information under circumstances where the researcher is confronted with elements that are too numerous – whose exact whereabouts are unknown, whose trajectories contain too many hiatuses, whose interrelations are yet to be fully grasped – which is often the case for the natural sciences such as physics and biology. That this condition of deficit should be turned into a universal goal of all scientific inquiry, wherein the researcher always keeps the phenomena of their research at a distance despite having privileged proximity to them (i.e., in inquiries into human society) – is a mistake that can be overcome by eclipsing the distance between the part and the whole in one single stroke, through the notion of the actor-network. The actor-network, as such, is not only a bridge between the micro and the macro, but also between the quantitative and the qualitative – in favour of the qualitative-quantitative.

ANT elected the sociology of Gabriel Tarde[7] as its chosen philosophical and algorithmic ancestor (Latour, 2010). Tarde engaged not only sociology, but also the human sciences more generally – history, geography, archaeology, social psychology, and economics. At the heart of Tarde's idea of the quantification of human society is that a structure cannot be qualitatively distinct from its components – only natural societies can, and even then it is because there is no way to do otherwise – and that the social sciences should not imitate the kind of quantification usually done in the natural sciences. Because human societies are accessible to researchers through their most intimate features, researchers of human societies should not let natural scientists dictate what their epistemology should be – but rather come up with their own definition of what it means for a discipline to be quantitative. Tarde's articulation of how quantities and qualities come about goes counter to how the narrative around the natural vs. the interpretive sciences is usually constructed: that the more preoccupied the researcher is with the individual, the local, and the situated, the more qualities will be made available, and that as we move away from the individual and towards the structural, we start to collect quantities. Tarde conceptualised the situation as the opposite: the more a researcher is engaged with the intimacy of the individual, the more discrete quantities they will end up with. It is only through moving away from the individual and creating distance that researchers begin to lose quantities – because we lack the instruments to collect enough quantitative evaluations (Latour, 2010). It is in this sense that the heart of all social phenomena is quantifiable, because 'individual monads are constantly evaluating one another in simultaneous attempts to expand and to stabilise their worlds' (Latour, 2010, p. 148). As such, for Tarde (and by extension, ANT), quantification begins with the individual and becomes difficult to maintain at the aggregate level – at which point qualities emerge as products of monadic interactions.

It is useful to think about diasporas as productive accomplishments of actors migrating, residing, coming back, creating bonds, maintaining ties, and forging connections – rather than an amorphous, broad, and analytically unsatisfactory aggregate that gestures at the phenomenon of population dispersion across time and space, that which shapes identities and cultures. We can even think of diasporas as processes of *diasporisation* (Olliff & Phillips, 2022), as ongoing societal transformation and development. Diasporic actor-networks are not displaced populations, whose attributes are determined on the individual level (gender, race, ethnicity, age, language, country of origin, etc.) before they are accounted for by means of enumeration – to determine quantities. Network accounts of diaspora are most compelling when they reveal relational dynamics among individuals and not their individual attributes; they focus on individuals not as members of discreet groups, but rather as members of overlapping networks. Diasporic actor-networks are made up of monads whose actual relations to each other do not preclude potentialities for transformation, change, or dissolution; it is only through the task of assembling that these networks come into existence.

How to assemble a diasporic network: formalist vs. relational approaches

So far, we have considered how the quantification of the individual is productive of the network, and how quantification does not necessarily deprive researchers of complexity and heterogeneity. We have also considered how ANT as a research paradigm enables a dynamic view of network that bypasses notions of structure and whole, and instead favours bundles of network ties between monads as always traceable and emerging. What ANT does not tell us, however, is *how* to assemble these ties. The task of network assemblage is left wide open; as long as researchers assemble their networks by 'following the actors' without ever shifting attention at any point to the idea of a whole, or changing modes of inquiry, then they can be said to have been faithful to the principles of ANT. In order to carry out the actual task of assembling networks, we now need to turn our attention to some theoretical approaches to analysing social networks. Chapter 2 will discuss a range of social network analysis techniques that can be applied to diaspora research.

Erikson (2013) observed that there are two distinct theoretical traditions of social network research: relationalism and formalism. The mixing of these logically inconsistent frameworks often gives the impression that social network research is atheoretical; however, Erikson (2013) argues that relationalism and formalism are internally consistent theoretical perspectives. Relationalism is a theoretical framework based on the primacy of relations rather than actor attributes; it rejects essentialism and *a priori* categories while emphasising the intersubjectivity of experience, the content of interaction among actors, as well as their historical settings. Formalism, on the other hand, hails from the theoretical works of Georg Simmel,[8] who were preoccupied with identifying *a priori* categories of relational types and patterns that are independent of cultural content or historical settings. As such, there are inherent tensions between these two approaches. These tensions do not necessarily mean that social network research employs 'a largely unorganised grab bag of measures, tools, and ideas' (Erikson, 2013, p. 3); rather, the contradictions existing between these two approaches mean that a unified framework for social network research cannot be built entirely on either foundation. At the same time, developing one unifying framework that encompasses both perspectives is unlikely. What's a researcher to do? Erikson (2013) suggests that any successful combination of the two approaches requires careful attention to the underlying presuppositions about perception, experience, and agency found in each. This section provides an overview of these approaches along with their underlying assumptions on context, meaning, and agency, and discusses the kinds of data that are most appropriate for different operationalisations of concepts and phenomena that are important to researchers of diaspora.

Formalism and social networks – the network flow approach

As a neo-Kantian, Simmel attempted to expand Kant's philosophy beyond the natural sciences to the social sciences. In *Critique of Pure Reason*, Kant argued that reason could produce new insights without recourse to empirical observation, for example, in geometry and arithmetic. Geometry could produce insights about the empirical world because the logic that made reasoning in geometry possible also made our experience of the world possible – rather than because geometry was modelled after the world. In this sense, knowledge that did not depend upon experience of the empirical world could nevertheless generate new insights. For example, the conception of space – argued Kant – is not only essential to geometry, but also necessary to our perception of the world. The correspondence between geometric space and perceptive space exists because the logic of geometry is part of the means by which we perceive the physical world. Because having the idea of space is necessary to experience the world, it must be prior to experience. As such, Kant's identification of the categories of meaning as prior to experience means that they must transcend any particular empirical setting as universal, fixed, and true – because they are part of the conditions that make experience possible. For Kant, essential structures of reason such as the ideas of time, space, and causality, are prior to our experience of the physical world; they make empirical observation possible.

If Kant asked, 'How is nature possible?' – then Simmel expanded and appropriated this question as 'How is society possible?' Following Kant, Simmel argues that this is 'answered by the conditions which reside *a priori* in the elements themselves, through which they combine, in reality, into the synthesis, society' (Simmel, 1971, p. 8). His research project, therefore, seeks to establish the social forms that constitute society as sociological apriorities; these social forms are not products of social relations that exist in the world, but rather ideal forms whose properties can be observed in relationships as experienced in the world. Simmelian social forms are discrete, pre-given, and fixed entities that exist outside of the material plane prior to their instantiation; they are not constitutive of relations, but rather give form to relations (Erikson, 2013). What would be the utility of the Simmelian social structural models of abstracted forms? Simmel provided many early examples of the impact of purely formal structures of relations between individuals as social network patterns. The most famous example is that of the triad, where Simmel pointed out how the shift from a dyad (a relationship between two people) to a triad (a relationship between three people) fundamentally changes both the nature of relationships between those actors and the potential for the kinds of patterns of social organisation that may occur (Erikson, 2018). While the dyad contains only two possibilities, which are either connection or disconnection, there are as many as 16 different possible configurations in a directed graph of three actors (see Figure 1.1). In a linked dyad, each dyadic actor has only one other link; as such, they are tightly bound to their partner. The transformation from a dyad to a triad alters the nature of the connection between individual actors: in a fully linked triad where all actors have relationships with each other, the intimacy of the relationship decreases and a sense of belonging to a group is created (Erikson, 2018).

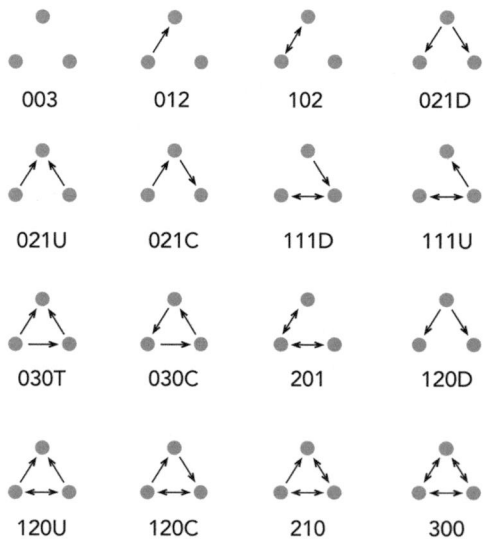

FIGURE 1.1 Directed triadic configurations.

The formal social form focuses on patterns of ties at the expense of the content of those ties; this basic tenet has encouraged a body of research into network dynamics that completely (and deliberately) ignore the content of social ties. Much research on the diffusion of innovation makes extensive use of this formalism; a paper on the diffusion of innovation among physicians by Coleman et al. (1957), for example, studied how relationships between doctors affected the pattern by which the pharmaceutical drug gammanym was adopted for use by different doctors. The study ignored altogether what gammanym does, the risks involved in its use, or its effectiveness – which is implied by its adoption. In the same vein, in a study across different domains of innovation (tetracycline, hybrid corn, family planning) by Valente (1996), variation in the adoption rates of these very distinct technologies is attributed solely to the social structures of the communities involved without considering the technologies' various affordances. This is because the process of innovation is understood as a social form through which the specific contents of different innovations pass (Erikson, 2018).

Social network research of the formalist flavour draws its analytical power from the ability to abstract away from details; it is in this sense that formalist network analysis is incompatible with the principles of ANT as discussed in the earlier section. Put in ANT terms, formalism is concerned with *intermediaries* rather than *mediators*. Intermediaries are transporters of meanings and can be metaphorically understood as pipes or vessels that do not transform the substances they carry: their input is their output. Understanding groups via intermediaries is similar to the *network flow approach* to social network analysis, where importance is placed on network paths of connection and disconnection. Network theories[9] falling under this approach include small-world theory, structural hole theory, Coleman social capital theory, and Grannovetter's strength of weak ties theory (Borgatti & Lopez-Kidwell, 2016). The principles of ANT, however, are more resonant – although not strictly aligned – with the relationalist agenda through its articulation of the mediator as a means to produce the social, as we will see below.

Relationalism and social networks – the network architecture approach

The fixity of formal social forms stands in contrast to the endless fluidity of a relationalist account of social life as open-ended, dynamic, and processual. Fuhse (2009), for example, articulates social networks as 'intersubjective constructs of expectations and cultural forms' (p. 52); all social networks are 'configurations of social relationships interwoven with meaning' (p. 51). For relationalists, meaning does not exist *a priori*, but rather emerges with and through interactions in co-constitutive ways. Erikson (2013) sharply remarks that another way of stating this approach would be to say that the meaning one individual assigns to another is the basis of any relationship; the absence of meaning can be interpreted as the absence of a relationship, in the sense that if one has no expectations or knowledge of another individual, then a relationship or connection cannot be said to exist. It is in this sense that meanings compose, rather than flow through, networks.

This collapse between meaning and network in relationalism resonates with ANT's collapse of the actor and the network; in fact, ANT can be understood as a form of relational materialism that codifies a body of ideas developed in the sociology and history of technology. In ANT's view, society is produced through the mutually constitutive interaction of a wide range of human and non-human entities. Relational sociologists share the view that transactions, social ties, and conversations constitute social life – in the sense that social life is *made up of*, rather than contains, connections and relations. In ANT terms, social life is made up of mediators, which 'transform, translate, distort, and modify the meaning or the elements they are supposed to carry' (Latour, 2005, p.39). Understanding groups via mediators is akin to the *network architecture approach* in social network analysis, where importance is placed on the coordination of network egos. Network success, according to this approach, is less about the transfer of resources (as in, network flow model), and more about the level of alignment between network nodes (Borgatti & Lopez-Kidwell, 2016). Within this approach, the mere existence of communication between nodes does not qualify for success terms – rather, attention is paid to the *content* of this communication. The content is also the main focus of the relational approach, where relations are never examined outside of their context.

Relationalism in sociology is usually positioned against substantialism – in which the source of social action and explanation in fixed entities exist independently of each other. These fixed entities can be individuals, societies, or social structures; when these fixed entities are said to interact, they are reified as if they were imbued with an existence of their own (Guy, 2018). The outcomes of these interactions exist in society, rather than constitute them. By contrast, concepts such as interaction, transaction, connection, and relation in the relationalist's arsenal of vocabulary call for a different way to think about social actors as mutually defining as they play their parts in a common process that unfolds over time – rather than disappear in it. Network analysis in the relational tradition does not aim to be transposable; it is meant to act as a useful analytical tool to make sense of particular settings. Indeed, network effects have been demonstrated to vary across contexts, providing a fruitful framework for researchers to conduct comparative studies in their relevant empirical domains.

This is not to say, however, that relational accounts of networks do not make a claim on generalisability – or universality – but rather do so with and through context. Formalist generalisability is often devoid of context and culture, as formal social forms are understood as crystallised, locally defined structures of social relations that constrain individual behaviours and network outcomes (Erikson, 2013). As such, even though formalism and relationalism are logically inconsistent, they are internally consistent. Each has a commitment to generalisability that goes beyond the universalism vs. particularism dichotomy, and there are alternative approaches within these two traditions that hold the potential for working through their contradictions (Erikson, 2013). Researchers of diaspora – particularly digital diaspora – should consider each approach in its full complexities even when they choose to align with either of the two approaches in their research practice. Frequently drawing on research from both approaches would enable researchers access to a large body of useful work across a wide variety of empirical domains. Erikson (2013) suggests that social network researchers could, apart from aligning themselves with one approach, also opt for professing theoretical pluralism (doing both), adopting a new theory powerful enough to encompass both approaches, and conducting empirical research exploring the areas of tension between the two research programs.

In practice, one way to combine the relational and formal traditions is the *network exchange* approach, which often involves experimentation where participants are placed into simulations of network environment for exchange. Their network positions are recorded and measured over time to discern the analytical usefulness of various network constructs in relation to each other, such as centrality over propagation[10] (Borgatti & Lopez-Kidwell, 2016). In these experiments, formal social forms and their effects are tested empirically (relationally), and their generalisability rests on statistical concepts such as sample size, sampling technique, representativeness, bias, replicability, and so on. These experiments, which usually take the form of network games designed so that participants can only negotiate certain communication goals with people they have been given ties to, precede the kind of massive experiments involving tens of millions of people on social media where the kind of information exposed to different test groups is manipulated. As people organise their social lives in an increasingly network-like fashion with digital technologies, it became possible to experiment on actual network processes by manipulating network variables that influence actual network outcomes – such as the examples with Facebook experiments in the 2010s discussed earlier in this chapter. It is also increasingly feasible to assemble diasporic networks with found digital data – such as social media behavioural data (Facebook 'likes', 'comments', 'reacts', 'shares'; Twitter 'retweets', 'loves', 'replies'; Instagram 'love', 'comments', 'shares') as records of interaction and exchange. Chapter 2 will discuss these issues in more detail, along with a case study that utilises Facebook behavioural data to assemble Vietnamese diasporic transnational networks.

Network research in the past decade has shifted towards utilising online interaction data generated as byproducts of everything we do on the internet to describe social processes through messy everyday interactions, experiment on network outcomes, and compare similar social processes across contexts. Even though techniques such as name generation and survey still retain their usefulness for various types of network research (cf. Pope & Lewis, 2008), their utility has become increasingly confined to narrow, pre-defined empirical domains. For network research that seeks to assemble open-ended social processes as ever-shifting interdependencies, digital data (web links, behavioural social media data, location data, Wikipedia contribution data, and so on) prove to be more appropriate. Research utilising historical records that contain granular network data such as family histories, financial records, or credit arrangements can be digitised and made to tell very compelling stories about the past with highly robust network concepts and measures (cf. Erikson, 2018). In the following section, we will explore how web data can be harnessed to assemble diasporic networks on the internet by considering the e-Diaspora Atlas project.

Case study: the e-Diasporas Atlas project

The e-Diasporas Atlas project (http://www.e-diasporas.fr/), developed and coordinated by Dana Diminescu at Télécom Paris, is among the few large-scale and continuing (at the time of writing) research projects that incorporate digital research methods into the study of diasporas in a systematic and concerted way. Some eighty researchers from diverse disciplines, laboratories, and countries took part in the project, producing thirty (and counting) interactive network maps of various e-diasporas by utilising curated web data. The e-diaspora concept as articulated in the context of this project is very specific; by mapping migrant websites created or managed by migrants and/or that deal with them, the project defines the e-diaspora as:

> a migrant collective that organises itself and is active first and foremost on the Web: its practices are those of a community whose interactions are 'enhanced' by digital exchange. An e-diaspora is also a dispersed collective, a heterogeneous entity whose existence rests on the elaboration of a common direction, a direction not defined once and for all but which is constantly renegotiated as the collective evolves. An e-diaspora is an unstable collective because it is redrawn by every newcomer. It is self-defined, as it grows or diminishes not by inclusion or exclusion of members, but through a voluntary process of individuals joining or leaving the collective – simply by establishing hyperlinks or removing them from websites. (e-diasporas, 2022, 'The Concept' section)

A distinction is made between concepts of the e-diaspora and the digital diaspora in this context. The e-diaspora is made up of networks of hyperlinks (website citation) as a proxy for diasporic activity – alongside content creation on these websites. A migrant website is defined as part of a given e-diaspora's occupation of the web; as such, it does not have to be located in a country away from the chosen country of origin. An e-diaspora is both online and offline in the sense that it presupposes knowledge of the diaspora in question and, based on exploration of the Web, reveals itself through the systematic mapping of these predefined web activities and exchanges. As such, the e-diaspora concept is very much a product of early web culture, where the emergence of e-diasporas can be understood as occurring in tandem with the diffusion of the internet and the development of multiple online public services (e-diasporas, 2022). The 'e-' prefix is reminiscent of the proliferation of web technologies in the late 1990s that captured the imagination of institutions, such as e-administration, e-democracy, e-education, e-healthcare, e-culture, and e-tourism. According to the project concept, the term e-diaspora is favoured over the term digital diaspora because:

> the latter may lend to confusion given the increasingly frequent use of the notions of 'digital native' and 'digital immigrant', in a 'generational' sense (distinguishing those born before from those born during/after the digital era). The object of the *e–Diasporas Atlas* is not this 'digital migrant', however, but the *connected migrant* [emphasis added] (e-diasporas, 2022, 'The Concept' section).

e-diasporas are, as such, relational networks assembled and disassembled as dynamic and ongoing processes of diasporisation. What's also interesting in this articulation of e-diaspora is its spatial implications – that e-diasporas *occupy* parts of the web by means of its proliferation. It is through this spatial conceptualisation that the methodology used to harmonise the various mappings of e-diasporas start with web exploration through a semi-automatic web crawler called Navicrawler – a Firefox add-on designed and developed by Mathieu Jacomy at Sciences Po Paris médialab. Through this crawler, researchers can identify the websites that a certain identified diasporic site links to (out-links), decide whether the sites are relevant to their research purpose, and label these sites for context. The researcher 'explores' the diasporic space in the same way that adventurers go on expeditions to discover unknown and unmapped terrains. The corpus resulting from this exploration exercise can be exported into a graph – which represents the websites as nodes, and edges as hyperlinks between them. In the second step of the research process, researchers can use a range of digital tools developed by the e-Diasporas team to retrieve general website information as registered with ICANN (such as geographical location, server hosting the website) from a list of URLs; text-mine from the index of the corpus a list of named entities such as persons, organisations, and places using Open Calais API; detect the languages used in each website as well as their distribution towards mapping multiculturalism.

The third step is to visualise the corpus collected and curated with Gephi – a general-purpose network visualisation software developed by Matthieu Bastian, Matthieu Jacomy, and Sébastian Heymann. For the purpose of the e-diaspora atlas, two types of visualisation are accommodated: spatialisation based on the physical principle of attraction/repulsion (according to the presence or absence of a link between two nodes) and geographical spatialisation using geocoded data. The last step entails populating the interactive (browsable) graphs of the corpus on the e-Diasporas project site, raw data used to generate these maps, and relevant statistics. The project site also hosts a range of working papers on diasporic network mapping by project researchers, spanning Breton, Chinese, Egyptian, Hmong, Indian, Italian, Jewish, Lebanese, Kerala, Macedonian, Mexican, Moroccan, Nepali, Palestinian, Russian, Sikh, Tamil, Tunisian, Turkish, Uyghur, Yugoslav, and Zoroastrian e-diasporas.

Following this overall four-step methodological strategy, project contributors were able to contextualise their own hyperlink network mapping to suit the particularities of the e-diaspora of their choice. Graziano (2012), for example, utilised the tools made available through the project to map a niche e-diaspora of highly skilled emigrants from Italy to compare its organisation with existing, known 'offline' Italian diasporic dynamics. With a sample of 159 websites (nodes) and 872 connections (edges), Graziano was able to show the extreme fragmentation of the Italian e-diaspora, which mirrors existing dynamics of the global Italian diaspora. Graziano (2012) attributed this fragmentation to the 'profound parochialism of a deeply regionalised country that was unified only 150 years ago' (p. 21). However, Graziano was also able to show the concentration of web presence in Sweden within this e-diaspora – which is reflective of the latest wave of migration by highly skilled ICT workers from Italy, which diverges from the general concentration of Italian diasporic communities in North America, South America, and Australia. In another study, Morgunova (2012) studied the Russian-speaking e-diaspora to discern the politics of belonging among Russian emigrants who position themselves as new minority groups in their host countries. Explicitly excluding Russian migrant websites written in German, French, English, Czech, and other languages, Morgunova made the analytical choice to explore the post-Soviet migrants who identify as Russian speakers and who wish to stay informed about discourse happening in 'mainland' Russia. The analysis also noted the absence of Russian women's professional networks while highlighting that diasporic Russian women were very active in organising community activities and language learning, campaigning for Russian language preservation, and advertising small women-led businesses across these migrant networks.

In the same series of papers, Reyhan (2012) explored the Uyghur e-diaspora to find that diasporic web content is highly political, while sites within the Uyghur region showed signs of self-censorship. Religious sites also proliferated within the e-diaspora; the case is the opposite within China, where they are poorly tolerated. The geographical grounding of the Uyghur e-diaspora is also quite interesting: while the largest concentration of websites is in the US and Turkey, France and Japan were neck and neck in third and fourth place. In both the US and Turkey, the majority of web content is in old Uyghur (compared to English and Turkish), whereas in Japan and France, the situation is the opposite: most content is produced in the 'host' languages of Japanese and French. Analysing the Palestinian e-diaspora across Germany, France, Italy, Austria, Australia, United States, Canada, Spain, Argentina, Chile, and Uruguay, Ben-David (2012) demonstrated a thematic and demographic shift from organisations of Palestinian communities abroad to a transnational solidarity network focused on Palestinian rights and the Boycott movement. Ben-David also showed, somewhat counter-intuitively, that the ties between diaspora and non-diaspora actors were stronger than among diaspora actors, which indicates that part of the dynamics of Palestinian communities is manifest not only between diaspora communities, but also between diaspora communities and civil society organisations in their host societies.

The e-diasporas atlas project is an example of how natively digital data can be collected with open-source tools and made to accommodate research questions by diaspora researchers. While hyperlink networks do not tell us everything there is to know about diasporic enactments as they pertain to the internet, they do provide interesting insights into how e-diasporas as web spaces might or might not reflect the dynamics of physical, non-web spaces where diasporic lives also unfold. It is also an example of how digital technologies have helped shape the agenda of diaspora research, both in the emerging techno-social enactments of diaspora that behove the attention of researchers as well as the tools and techniques that are becoming available to social research. The profound ways in which digital connectivity has transformed spatiality, belonging, and self-identification generate exciting opportunities for researchers to study the migrant experience on a much larger scale than possible before while assembling and analysing complex patterns of globalisation and localisation. Just as the notion of diaspora gestures towards a post-national space that problematises the relationship between nation, soil, and identity (Ponzanesi, 2020), the network concept opens up the task of tracing the processes of diasporic formation as dynamic, hybrid, and open-ended – thereby problematising the concept of diaspora in all its complex diversity and fleeting changeability.

Conclusion

This chapter has introduced the multidisciplinary paradigm of network analysis and discussed how network analysis can be embraced for the qualitative-quantitative study of diasporas. There are a few analytical and theoretical implications to conceptualising diasporas as quantifiable networks. Firstly, this approach is explicitly topographical in nature and is not equipped to account for the affective, lived dimensions of the migrant experience. In order to explore the lived experiences of migrants, researchers should turn to approaches such as digital ethnography and participant observation, where the locality, mobility, and digitality of the migrant experience can be foregrounded. These approaches are discussed in Chapters 4 and 5 of this book.

Secondly, networks are qualitatively distinct from communities – a concept that has been at the heart of the conceptualisation of diasporas since Benedict Anderson's seminal work in 1983. In an oft-quoted definition, Anderson (1991) remarked that the nation as a community 'is imagined because the members of the smallest nation will never know most of their fellow-members, meet them or even hear of them, yet in the minds of each lives the image of their communion' (p. 6). Communities are also not to be distinguished by their 'falsity' or 'genuineness', according to Anderson, but by the ways in which they are variously imagined. A community, as such, is larger than the sum of its parts; insofar as it is 'imagined', it also defies the kind of empirical tracing done by relationalists and ANT researchers. In reconfiguring diasporas as networks, researchers make the conceptual shift to operationalise diasporas as bundles of associations, relations, interactions, transactions – and nothing else. What a network approach offers are analytical pathways for the mapping of the cartographies of diasporic selves in relation to ongoing processes of globalisation and localisation. In the next chapter, we will explore a range of social network analysis techniques and measures that are useful to researchers of diaspora.

Notes

1 In 1736, Leonhard Euler – a Swiss mathematician – used the graph method to prove the impossibility of a single path that crosses all seven bridges in the city of Königsberg in a paper that is often cited as the earliest paper in both topology and graph theory. Widely recognised as the originator of modern graph theory, Euler's study of graph theory seems to have been entirely grounded in symbolic representations rather than in graph drawing. We will return to the relationship between visualisation and the generation of insights using graphs in Chapter 2.

2 More than a century after Euler's application of the graph method, Arthur Cayley – a British mathematician – became interested in the tree – a class of graph that is undirected, and in which any two vertices are connected by exactly one path. Cayley coined the concept of the tree as an analytical form arising out of differential calculus in 1857. We will discuss directed and undirected graphs in Chapter 2.

3 This is the first in a series of eight papers published by Paul Erdős and Alfréd Rényi between 1959–1968, setting the tone for network research in the decades to come. For a list of these papers, refer to the Bibliography.

4 American sociologist Mark S. Granovetter demonstrated in this paper that weak ties are important in creating cohesion between groups, especially in the contexts of diffusion of information and influence, mobility opportunity, and community organisation. Social network analysis is often deployed as a tool for linking micro and macro levels of sociological theory – this is something we will discuss in Chapter 2.

5 We will discuss these concepts in Chapter 2.

6 ANT originated in science studies in the 1980s and became influential in the social sciences more broadly in the late 1990s. ANT is known for its iconoclastic style of argument – a source of inspiration and criticism in equal measures. Ethnomethodologist Michael Lynch (1995, p. 168) mockingly suggested that a more appropriate name for ANT might be 'actant-rhizome ontology' – a suggestion that Bruno Latour, in classic ANT fashion, fully embraced. Latour's reservation with this moniker is, however, that it is a 'horrible mouthful of words – not to mention the acronym ARO' (Latour, 1999, p.19).

7 Gabriel Tarde (1843–1904) was a French sociologist, criminologist, and social scientist whose most well-known work is perhaps *Les Lois de l'imitation* (1890; *The Laws of Imitation*), in which he articulated the notion of imitation as mould of social relations through an interspsychology that makes all communication possible. This notion of imitation is also at the

heart of the famous Tarde vs. Durkheim debate on the nature of sociology. Tarde argued that society is made up of individuals, and that the social psychology of their interaction brings about social structures and change. Durkheim argued that sociology should be conceptualised on a level of its own – one that avoids reduction to individual-level psychology – and instead focused on norms that constrain behaviour as external to the individual. In Tarde's view, norms are products of interaction and therefore not exterior to individuals.

8 Georg Simmel (1858–1918) was a German sociologist and neo-Kantian philosopher, best known for his works on sociological methodology. Simmel sought to isolate the general or recurring forms of social interaction from the more specific kinds of activity, such as political, economic, and aesthetic. He also outlined a range of distinctive concepts of contemporary sociology, such as social distance, marginality, urbanism as a way of life, roleplaying, social behaviour as exchange, conflict as an integrating process, dyadic encounter, circular interaction, reference groups as perspectives, and sociological ambivalence.

9 These will be discussed in detail in Chapter 2.

10 We will explore these concepts in Chapter 2.

Bibliography

Anderson, B. (1991). *Imagined communities: Reflections on the origin and spread of nationalism.* Verso Books.

Baiocchi, G., Graizbord, D., & Rodríguez-Muñiz, M. (2013). Actor-network theory and the ethnographic imagination: An exercise in translation. *Qualitative Sociology, 36*(4), 323–341.

Barabási, A.-L. (2016). *Network science.* Cambridge University Press, UK.

Ben-David, A. (2012). The Palestinian diaspora on the Web: Between de-territorialization and re-territorialization. *e-Diasporas Atlas.* Available at http://www.e-diasporas.fr/wp/ben-david.html.

Bond, R. M., Fariss, C. J., Jones, J. J., Kramer, A. D., Marlow, C., Settle, J. E., & Fowler, J. H. (2012). A 61-million-person experiment in social influence and political mobilization. *Nature, 489*(7415), 295–298.

Borgatti, S. P., & Lopez-Kidwell, V. (2016). Network theory. In J Scott, & P J Carrington (Eds.), *The SAGE handbook of social network analysis.* EBook Central, ProQuest.

Cayley, A. (1857). On the theory of the analytical forms called trees. *Philosophical Magazine, Series IV, 13*(85), 172–176. 10.1017/CBO9780511703690.046.

Coleman, J., Katz, E., & Menzel, H. (1957). The diffusion of an innovation among physicians. *Sociometry, 20*(4), 253–270.

e-diasporas (2022). The concept. The e-Diaspora Atlas project. Available at http://www.e-diasporas.fr/#about.

Erdős, P. (1962). On the number of complete subgraphs contained in certain graphs. *Magyar Tud. Akad. Mat. Kutató Int. Közl, 7*(3), 459–464.

Erdös, P. (1963). On the structure of linear graphs. *Israel Journal of Mathematics, 1*(3), 156–160.

Erdos, P. (1964). On the number of triangles contained in certain graphs. *Canad. Math. Bull, 7*(1), 53–56.

Erdős, P., & Rényi, E. (1959). On random graphs I. *Publicationes Mathematicate, 6*, 290–297.

Erdos, P., & Rényi, A. (1960). On the evolution of random graphs. *Publ. Math. Inst. Hung. Acad. Sci, 5*(1), 17–60.

Erdős, P., & Rényi, A. (1961). On the strength of connectedness of a random graph. *Acta Mathematica Hungarica, 12*(1), 261–267.

Erdős, P., & Rényi, A. (1966). On the existence of a factor of degree one of a connected random graph. *Acta Math. Acad. Sci. Hungar, 17*, 359–368.

Erdős, P., & Rényi, A. (1968). On random matrices II. *Studia Sci. Math. Hungar, 3*, 459–464.

Erikson, E. (2013). Formalist and relationalist theory in social network analysis. *Sociological Theory, 31*(3), 219–242.

Erikson, E. (2014). *Between monopoly and free trade: The English East India Company, 1600–1757.* Princeton University Press, USA.

Erikson, E. (2018). Relationalism and social networks. In F Dépelteau (Ed.), *The Palgrave handbook of relational sociology* (pp. 271–287). Palgrave Macmillan, Cham.

Euler, L. (1736). Solutio problematis ad geometriam situs pertinentis (The solution of a problem relating to the geometry of position). *Commentarii Academiae Scientiarum Petropolitanae*, 128–140.

Fuhse, J. A. (2009). The meaning structure of social networks. *Sociological Theory*, 27(1), 51–73.

Granovetter, M. S. (1973). The strength of weak ties. *American Journal of Sociology*, 78(6), 1360–1380.

Graziano, T. (2012). The Italian e-Diaspora: Patterns and practices of the Web. e-Diasporas Atlas. Available at http://www.e-diasporas.fr/wp/graziano-italian.html.

Guy, J. S. (2018). Is Niklas Luhmann a relational sociologist? In F Dépelteau (Ed.), *The Palgrave handbook of relational sociology* (pp. 289–304). Palgrave Macmillan, Cham.

Jones, J. J., Bond, R. M., Bakshy, E., Eckles, D., & Fowler, J. H. (2017). Social influence and political mobilization: Further evidence from a randomized experiment in the 2012 US presidential election. *PloS One*, 12(4), e0173851.

Kramer, A. D., Guillory, J. E., & Hancock, J. T. (2014). Experimental evidence of massive-scale emotional contagion through social networks. *Proceedings of the National Academy of Sciences*, 111(24), 8788–8790.

Latour, B. (2005). *Reassembling the social: An introduction to actor-network-theory*. Oxford University Press.

Latour, B. (2010). Tarde's idea of quantification. In M Candea (Ed.), *The Social after gabriel tarde: Debates and assessments* (pp. 161–178). Routledge.

Latour, B. (1999). On recalling ANT. In J Law, & J Hassard (Eds.), *Actor-network theory and after* (pp. 15–25). Blackwell Publishers, Oxford.

Law, J. (1999). After ANT: complexity, naming and topology. *The Sociological Review*, 47(S1), 1–14.

Lury, C. (2021). *Problem spaces: How and why methodology matters*. John Wiley & Sons.

Lynch, M. (1995). Building a global infrastructure. *Studies in History and Philosophy of Science*, 26(1), 167–172.

Mol, A. (2008). *The logic of care: Health and the problem of patient choice*. Routledge.

Morgunova, O. (2012). National Living On-Line? Some aspects of the Russophone e-diaspora map. *e-Diasporas Atlas*. Available at http://www.e-diasporas.fr/wp/morgunova.html.

Olliff, L. & Phillips, M. (2022). Introduction. In M Phillips, & L. Olliff (Eds.), *Understanding diaspora development: Lessons from Australia and the Pacific*. Palgrave.

Ponzanesi, S. (2020). Digital diasporas: Postcoloniality, media and affect. *Interventions*, 22(8), 977–993.

Pope, J., & Lewis, J. M. (2008). Improving partnership governance: Using a network approach to evaluate partnerships in Victoria. *Australian Journal of Public Administration*, 67(4), 443–456.

Reyhan, D. (2012). Uyghur diaspora and Internet. *e-Diasporas Atlas*. Available http://www.e-diasporas.fr/wp/reyhan.html.

Simmel, G. (1971). *On individuality and social forms*. D N Levine (Ed.). University of Chicago Press, Chicago.

Valente, T. W. (1996). Social network thresholds in the diffusion of innovations. *Social Networks*, 18(1), 69–89.

2

SOCIAL NETWORK ANALYSIS AND THE DIASPORA

Introduction

Social network analysis is a highly formal approach to representing and investigating social relations. All social networks are contextual in the sense that they are formalised representations of certain aspects of social phenomena; these particular aspects are always specific to context and research purpose as they become represented in ways that make them amenable to quantitative exploration. Hennig et al. (2013) remark that when we say we are studying social networks, what we really mean to say is that we are studying social phenomena *by means of* network representations: 'gathering data about aspects of a phenomenon and organising the data in a convenient form, ... applying methods that produce additional, derived data, and translating these back to the realm of the phenomenon' (p. 15). Much like other empirical inquiries, social network analysis entails a level of abstraction whose usefulness depends on how well the local operationalisation of network concepts faithfully translate, illuminate, and mobilise underlying social phenomena. Social network analysis as a method has its own specific set of assumptions and logic that motivate network representations through data collection exercises that capture social phenomena in certain lights but not others; just as surveys limit the phenomena being examined to the specific questions asked, network data operationalise and enact the phenomena they seek to study through formalised quantification of pair-wise relationships.

Social network analysis is one of many ways with which researchers of diaspora can conceptualise, analyse, and assemble the social phenomenon of their preoccupation: that of the ever-evolving diasporic network. Much like how tolerance, acquaintanceship, friendship, or kinship can be operationalised in network terms as a series of interactions and non-interactions, diasporic networks can similarly be understood in terms of the absence or existence of traceable relations in context. A network account of diaspora can help researchers explore the constraining and enabling structural qualities of diasporic networks, observe patterns of assembly and

DOI: 10.4324/9781003336556-4

disassembly, and even propose network interventions towards the maintenance and production of strong, lasting diasporic networks. In the next section, we consider what constitutes a network in social network analysis – its underlying assumptions and justifications, the data required for an empirical network study, and the ways in which networks can be represented.

Of nodes and ties: what's in a network?

Networks are highly graphic: images of points and connecting lines forming structures of varying degrees of complexity often come to mind whenever the concept is evoked. In its simplest definition, a network is a collection of connected objects, where the objects are graphically represented as nodes or vertices (drawn as points) and their connections represented as edges (drawn as lines between points). That networks are often thought about as graphs is a relatively new phenomenon, following shortly after – but not coinciding entirely with – the invention of graph theory in Leonhard Euler's 1736 paper, where he solved the path-tracing problem posed by the bridges of Königsberg entirely through symbolic representations, i.e., without drawing any graphs (see Chapter 1). This is to say that we can have networks without graphs and vice versa – just as networks are among the ways in which social phenomena can be expressed, graphs are also only among the many ways in which networks can be represented. Graphical visualisation was driven historically by pedagogy, exposition, record keeping, and mathematical recreation (Kruja et al., 2002); that graphical drawing became a locus of important innovation – particularly in the development of visual aids for solving mathematical problems – is seen as a historical curiosity, given how mathematical recreations are typically seen as not having significant contributions to the field of mathematics (Kruja et al., 2002). Many insights can be gained without graphical visualisation; in fact, in the case of highly complex networks, it is often very difficult to discern useful insights from examining their visualisations, which can be done relatively quickly with the aid of computer programs.

Consider the undirected and directed graphs in Figure 2.1. The number of nodes is identical; there are more edges in the directed graph (G_2), and the edges represent the direction of the relations between nodes in addition to the existence of a relation. We can tell from both graphs that A is connected to C, E, and F; in (G_2), A is connected to C and F in both directions. Alternatively, we can also say that in (G_2), C, E, and F are reachable from A, and A is reachable from C and F. Graph drawing is not the only way we can represent these relationships, however. We can construct an adjacency matrix of Booleans (0's and 1's), where the Boolean value indicates if there is a direct path between two nodes. M_1 represents (G_1) in this way: if there is a path between two nodes, the value of their corresponding cell is 1, otherwise, the value is 0. In the case of (G_2), where the edges are directed and two nodes are connected by more than one edge, the corresponding adjacency matrix is not symmetric, since the existence of a directed edge from one node to another does not necessarily imply the existence of a directed edge in the reverse direction. M_2 represents (G_2) as an asymmetric adjacency matrix, where each entry value is the number of directed edges from one node to another – read from left to right.

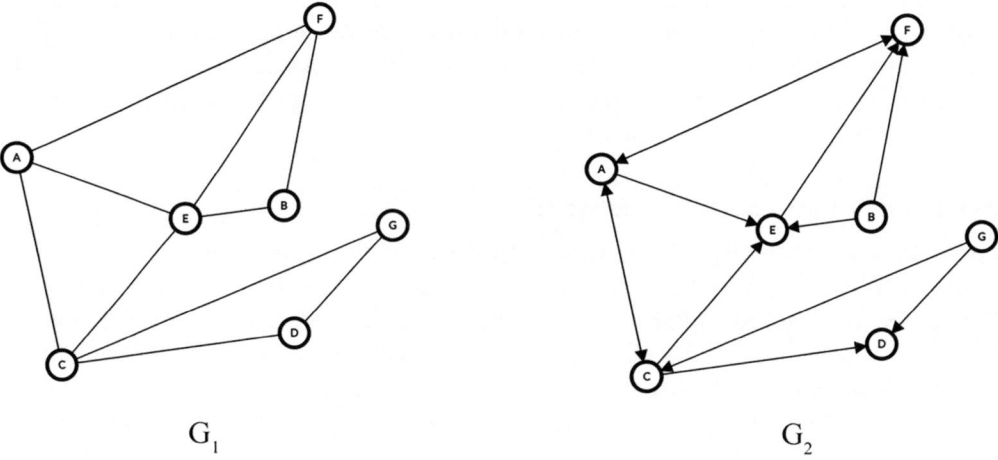

$$G_1 \qquad\qquad G_2$$

FIGURE 2.1 Undirected and directed graphs.

$$M_1 = \begin{bmatrix} 0 & 0 & 1 & 0 & 1 & 1 & 0 \\ 0 & 0 & 0 & 0 & 1 & 1 & 0 \\ 1 & 0 & 0 & 1 & 1 & 0 & 1 \\ 0 & 0 & 1 & 0 & 0 & 0 & 1 \\ 1 & 1 & 1 & 0 & 0 & 1 & 0 \\ 1 & 1 & 0 & 0 & 1 & 0 & 0 \\ 0 & 0 & 1 & 1 & 0 & 0 & 0 \end{bmatrix} M_2 = \begin{bmatrix} 0 & 0 & 1 & 0 & 1 & 1 & 0 \\ 0 & 0 & 0 & 0 & 1 & 1 & 0 \\ 1 & 0 & 0 & 1 & 1 & 0 & 0 \\ 0 & 0 & 0 & 0 & 0 & 0 & 0 \\ 0 & 0 & 0 & 0 & 0 & 1 & 0 \\ 1 & 0 & 0 & 0 & 0 & 0 & 0 \\ 0 & 0 & 1 & 1 & 0 & 0 & 0 \end{bmatrix}$$

Another way to represent (G_1) and (G_2) is with adjacency lists. With adjacency lists, each node stores a list of its adjacent vertices; for undirected graphs, each edge from i to j would be stored twice: once in i's list of neighbours and once in j's list of neighbours. Below are how (G_1) and (G_2) are represented as adjacency lists in the Python programming language:

```
G₁ = {'A' : set(['C', 'E', 'F']),
'B' : set(['E', 'F']),
'C' : set(['A','D', 'E', 'G']),
'D' : set(['C', 'G']),
'E' : set(['A', 'B', 'C', 'F']),
'F' : set(['A','B', 'E'])
'G' : set(['C', 'D'])}

G₂ = {'A' : set(['C', 'E', 'F']),
'B' : set(['E', 'F']),
'C' : set(['A','D','E']),
'D' : set(),
'E' : set(['F']),
'F' : set(['A']),
'G' : set(['C', 'D'])}
```

Each of these methods of representing networks has its own comparative advantages and disadvantages. When working with a computer, the adjacency matrix representation takes constant time (just one memory access) to determine whether or not there is an edge between any two given vertices. The most important disadvantage of the adjacency matrix representation is that it requires n^2 storage, even if the graph has as few as $O(n)$ edges; just examining all the entries of the matrix would require n^2 steps, thus precluding the possibility of linear time algorithms for graphs with many fewer than n^2 edges (Wagner, 2003). Adjacency lists avoid these disadvantages: the total storage used by an adjacency list representation of a graph with n vertices and m edges is $O(n + m)$, making it appropriate for graph algorithms that take linear or near linear time. One potential disadvantage of adjacency lists is that determining whether there is an edge from node i to node j may take as many as n steps since there is no systematic shortcut to scanning the adjacency list of node i (Wagner, 2003). Gephi – a free network visualisation tool used in the case study included in this chapter – allows you to work with simple edge lists, adjacency matrices, or adjacency lists to generate network graphs.

Networks can also be weighted or unweighted. A weighted network is a network where the ties among nodes have weights assigned to them; in popular network visualisation, this is often represented by the thickness of ties. When represented on adjacency matrices, the value of [i, j]th cell can be larger than 1. The weight of ties is usually used to represent the frequency of interaction between two nodes, where interaction is the translation between a given social concept (e.g., friendship) and its empirical operationalisations (e.g., frequency of node exposure, time exposed, timing of interaction).

Why are we interested in social networks as representations of relational social phenomena? Social network as understood in this chapter is much narrower than the kind of network articulated in, for example, Actor-Network Theory (see Chapter 1) – where all sorts of associations are understood as part of an actor-network (between a person and a computer, a chair, a colleague, an ideology). Social network analysis is a more stringent approach to uncovering social structures that can limit or enable nodes in the network based on how the network is connected (Hogan, 2017). Edges in a social network analysis dataset are always roughly comparable – such as email correspondence, friend connections on Facebook, threaded conversations on a Facebook diasporic community page, or webpage links. By using social network analysis techniques, we seek to understand social dynamics not through individual attributes (i.e., individuals as members of discreet groups), but through the ways in which they are connected (i.e., individuals as members of overlapping networks).

Social networks can be understood roughly as either simple or complex. Hogan (2017) defines simple networks as those that consider nodes as discrete entities and edges as some sort of association between these nodes. Complex networks, on the other hand, are networks that might signify more than one type of relation between nodes, include some sort of time dimension (temporal networks), or feature multiple types of nodes. Simple networks can be categorised into three levels: personal networks, partial networks, and whole networks (Hogan, 2017). Personal networks are made up of personal connections ('alters') that are linked to a focal individual ('ego'); a network that contains only alters who are directly connected to an ego is called a 1-degree network. A network that also contains the connections among alters is colloquially referred to as a 1.5-degree network (Hogan, 2017). Partial networks look beyond the connections around ego – such as discourse networks within a self-organising Facebook community, or the diffusion of a Twitter hashtag. Partial networks allow us to analyse the social structure of groups that organise within specific domains. Whole networks are networks that have meaningful boundaries (Hogan, 2017), where we can reasonably articulate why certain nodes and edges should be included and not others. Many whole networks, as such, can add up to partial networks. In the case study included at the end of this chapter, many discrete whole networks of traditional medicine groups that organise around particular non-biomedical practices on Facebook add up to a partial discourse network of Vietnamese non-biomedical medical practices.

Complex networks can be roughly categorised into multiplex networks, temporal networks, and modal networks (Hogan, 2017). Multiplex networks contain interlayer edges that connect nodes representing the same actor in different layers. As such, multiplex networks typically represent different sets of interactions between the same (or a similar set) entities (e.g., individuals, organisations). Temporal networks are networks whose edges are active only at certain points in time; these networks are time-varying in that they are characterised by intermittent activation of individual connections between nodes. Modal networks, or multimodal networks, are heterogeneous networks where each node belongs to a particular mode, and edges belong to a particular cross net – understood as a particular kind of interaction between two modes. Nodes are never linked directly but by their shared modal association. Complex networks as introduced in this chapter are networks of *multiple types*: of nodes, edges, or temporality. In the next section, we will consider some basic network statistics that are useful to the quantitative analysis of network structures as enabling and constraining mechanisms of social relations.

Basic network statistics and concepts

Basic network statistics and concepts can be organised into five levels: monadic (position of a specific node), dyadic (two nodes), triadic (three nodes), meso (clusters of nodes), and the level of the whole network (Hogan, 2017). In practice, we usually conduct analysis across these different levels to examine different parts of an overall research question. This section introduces some of the most frequently used statistics and concepts in social network analysis and their applications in the social sciences.

Monadic analysis

When analysing the position of a node, we often work with the concept of centrality – understood broadly as how 'central' a node is within its network. There are many ways we could calculate a node's centrality. *Degree centrality* is the total number of nodes adjacent to a given node, which can be calculated by either a raw count or normalised by dividing the total number of nodes in a network – in which case it would take up a value between 0 and 1. Degree centrality allows researchers to get a broad sense of which nodes are most well-connected. A more specific version of this statistic is *in/out-degree centrality*, which counts only edges either going into or out from a node. While degree centrality gives us a crude overview of node centrality, it fails to give us a nuanced account of the centrality concept: not all connections contribute equally to a node's centrality, as the centrality of adjacent nodes also matter: a node might be positionally better off to have few connections to a few very central nodes than to have a lot of connections to poorly connected nodes. *Eigenvector centrality* is a statistic that considers the centrality of adjacent nodes; for a given graph $G = (V, E)$ with $|V|$ nodes, let $A = (a_{v,t})$ be the adjacency matrix ($a_{v,t} = 1$ if node v is connected to node t, and $a_{v,t} = 0$ otherwise). The centrality score x_v of node v is defined as:

$$x_v = \frac{1}{\lambda} \sum_{t \in M(v)} x_t = \frac{1}{\lambda} \sum_{t \in G} a_{v,t} x_t$$

Where $M(v)$ is the set of neighbours of v, and λ is a constant. In general, there will be many different eigenvalues λ for which a non-zero eigenvector solution exists. In practice, many network analysis programs allow researchers to calculate the eigenvector centrality score easily; this score is interpreted as the measure of influence a node has on a network – the higher the score, the more influence the node has. In economic public goods problems, for example, a person's eigenvector centrality has been used to understand how much that person's preferences influence an efficient social outcome (Elliott & Golub, 2013). In diaspora research, an actor's eigenvector centrality score could also potentially be used to understand their influence on certain collective outcomes within the diasporic network.

There are also centrality measures that quantify node importance with regard to assumptions other than how well-connected adjacent nodes are, such as closeness centrality and betweenness centrality. *Closeness centrality* measures how easy it is for a node to reach all other nodes in a network; it is calculated as the average of the shortest path length from the node to every other node in the network (Freeman, 1978/79). Formulated this way, closeness centrality pinpoints well-connected nodes that may reach any other nodes within a few hops, as well as less connected nodes that may be very distant in the graph. A lower closeness score translates to more centrality; for example, if node A has a closeness centrality of 2.3 and node B has a closeness centrality of 3.4, node A is considered more central by this measure. The benefits of closeness centrality are that it indicates nodes as more central if they are closer to most of the nodes in the graph. This strongly corresponds to visual centrality—a node that would appear towards the centre of a graph when we draw it usually has a high closeness centrality. This is very different from eigenvector centrality, which assumes that having important contacts – rather than having more contacts – is the main criterion of connectedness. With eigenvector centrality, the centrality of a node is related to the sum of the centrality of its neighbours.

Betweenness centrality indicates how much a node is 'between' others – it is conventionally calculated as the fraction of shortest paths between node pairs that pass through the node of interest (Freeman, 1978/79). Similar to closeness centrality, betweenness centrality measures the influence a node has over the diffusion of information through the network. Given $G = (V, E)$ and s, t are a fixed pair of graph nodes, the betweenness score of a node v for this node pair is the fraction of shortest paths between s and t that include v. The betweenness centrality of v is the sum of its betweenness scores for all possible pairs of s and t in the graph. Let σ_{st} be the number of shortest paths between s and t, and let $\sigma_{st}(v)$ be the number of those shortest paths that pass through v. The betweenness centrality of node v is defined as:

$$G(v) = \sum_{s \neq v \neq t} \frac{\sigma_{st}(v)}{\sigma_{st}}$$

In practice, betweenness centrality is also easily calculated with popular social network analysis programs such as Gephi – which calculates betweenness centrality conventionally as the number of shortest paths between node pairs that pass through that node. By conventionally calculating the shortest paths between pairs of nodes, however, we assume that information spreads only along those shortest paths. A variant of this measure uses random walks, counting how often a node is traversed by a random walk between two other nodes (Newman, 2005). This variant relaxes the assumption made in the conventional calculation by including contributions from essentially all paths between nodes rather than just the shortest while still giving more weight to short paths. Betweenness centrality measures the extent to which a node plays a bridging role in a network; a node with high betweenness is likely to be aware of what is going on in multiple social circles. Individuals with high betweenness centrality in diasporic networks tend to be individuals on whom people within their various social circles rely to make connections with other people. Identifying these individuals could be useful for qualitative interviews that would augment our understanding of connections between network clusters, as well as insights into what connects them and what keeps them separate. 'Betweenness central' individuals may act as brokers of structural holes – 'empty spaces' between contacts in a person's network who do not interact closely, even though they may be aware of one another (Labun & Wittek, 2014). They may also act as gatekeepers between these isolated clusters as structural holes reflect 'an opportunity to broker the flow of information between people, and control the projects that bring together people from opposite sides of the hole' (Burt, 2000, p. 353).

Finally, the *local clustering coefficient* is a measure of the degree to which an ego's peers in a graph tend to cluster together. This measure is crucial to the development of the small world hypothesis (Watts & Strogatz, 1998), which is discussed in Chapter 3's case study. The local clustering coefficient of a node quantifies how close its neighbours are to being a clique within the whole network; this measure gives an indication of the embeddedness of single nodes. Given $G = (V, E)$, and e_{ij} connects v_i and v_j, the neighbourhood N_i for v_i is defined as its immediately connected neighbours as:

$$N_i = \{v_j : e_{ij} \in E \lor e_{ji} \in E\}$$

Let k_i be the number of nodes in N_i, the local clustering coefficient C_i for v_i is given by a proportion of the number of links between the nodes within its neighbourhood divided by the number of links that could possibly exist between them. The local clustering coefficient for undirected networks is defined as:

$$C_i = \frac{2|\{e_{jk} : v_j, v_k \in N_i, e_{jk} \in E\}|}{k_i(k_i - 1)}$$

The local clustering coefficient for directed networks is calculated a bit differently since e_{ij} is distinct from e_{ji}. In directed networks, there are $k_i(k_i - 1)$ edges that could exist among the nodes within N_i – whereas in undirected networks, there are $\frac{k_i(k_i - 1)}{2}$ edges in N_i, since e_{ij} and e_{ji} are identical. The local clustering coefficient for directed networks is therefore defined as:

$$C_i = \frac{|\{e_{jk}: v_j, v_k \in N_i, e_{jk} \in E\}|}{k_i(k_i - 1)}$$

In most real-world networks, particularly social networks, nodes tend to create tightly knit groups characterised by a relatively high density of ties; this likelihood tends to be greater than the average probability of a tie randomly established between two nodes (Holland & Leinhardt, 1971; Watts & Strogatz, 1998). On average, nodes of higher degree centrality exhibit lower local clustering; having a lower local clustering coefficient means that a node is better embedded in the whole network and as such has more influence. The lower a node i's local clustering coefficient, the more structural holes there are in the network around i. It is in this sense that i has power over information flows between these otherwise non-overlapping clusters.

Dyadic analysis

In dyadic analysis, we are interested in how two nodes relate to each other. There are two main concepts that we work with in this regard: reciprocity (whether a connection goes both ways), and homophily (whether individuals of the same type are particularly prone to connect to each other). Reciprocity is variously operationalised across different empirical domains. In a purchase situation, for example, one party provides goods or services, and the other reciprocates with payment in currency or in-kind. On social media, the factors underlying the reciprocal sharing of information and social support are more abstract reciprocal exchanges; reciprocity can be operationalised as 'following back' behaviour on bidirectional rating and review systems including Uber, Lyft, Airbnb, eBay, and Couchsurfing (Starr Jr. et al., 2020). The concept of *guanxi* – operationalised in the sociology literature as a social network of reciprocal personal connections – has been extensively studied and is considered central to personal and business relationships in China (Park & Lou, 2001). Reciprocity is at the heart of network exchange theories, which theorise reciprocity as a behavioural response to perceived kindness and unkindness, where kindness comprises both distributional fairness as well as fairness intentions (Molm et al., 2007). In game theory literature, Falk and Fischbacher (2006) present a formal theory of reciprocity, which takes into account that people evaluate the kindness of an action not only by its consequences but also by its underlying intention. Empirical studies on network reciprocity are usually conducted to either confirm or dispute aspects of social exchange theory. A study by Surma (2016), for example, found strong evidence that an increase in the number of reciprocity messages an ego broadcasts in online social networks increases the reciprocity reactions from the network – which appears to be congruent with the stipulations of social exchange theory. Researchers of diaspora can similarly explore whether social exchange theory holds up in various diasporic contexts by examining reciprocal practices among members of diasporic networks.

Homophily as a network principle states quite simply that similarity breeds connection: homophily structures network ties of every type, including marriage, friendship, work, advice, support, information transfer, exchange, comembership, and other types of relationship (McPherson et al., 2001). Homophily limits people's social worlds in a way that has powerful implications for the information they receive, the attitudes they form, and the interactions they experience; homophily in race and ethnicity creates the strongest divides in our personal environments, with age, religion, education, occupation, and gender (McPherson et al., 2001). Contact between similar people occurs at a higher rate than among dissimilar people; ties between non-similar individuals also dissolve at a higher rate, which sets the stage for the formation of niches (localised positions) within social networks. Homophily implies that distance in social characteristics translates into distance in network; it also implies that any social entity – be it attitude, knowledge, sentiment, or ideology – that depends to a substantial degree on networks for its transmission will tend to be localised in social space and will obey certain fundamental dynamics as it interacts with other social entities in an ecology of social forms (McPherson et al., 2001).

Homophily as a concept has received significant purchase in network research over the past decade, especially within the online social network literature. Blex and Yasseri (2022), for example, show how under the assumption of homophily, echo chambers, and fragmentation are system-immanent phenomena of highly flexible social networks, even under ideal conditions for heterogeneity. They also argue that no level of positive algorithmic bias in the form of rewiring is capable of preventing fragmentation and its effect on reducing the fragmentation speed is negligible. Dinh et al. (2022) study the evolution of online dating through large-scale data analysis of the online dating site eharmony over a decade and find that similarity between profiles – an operationalisation of homophily – is overall not a predictor for courting success, except for similarity in the number of children and smoking habits. Laniado et al. (2016) explored gender homophily – defined as a preference for interactions with individuals of the same gender – by analysing the interactions of 10 million users of Tuenti, a Spanish social networking service popular among teenagers. They found that in dyadic relationships, gender homophily is higher among women – whereas there is little or no gender homophily among men. However, when examining the gender composition of triangle motifs, they observe a marked tendency of users to group into gender-homogeneous clusters, with a particularly high number of male-only triangles. These dynamics also differ across age, with higher homophily among teenagers in both dyadic and triadic relationships. Examining homophily in diasporic networks could have interesting implications: if we understand diasporic clusters as homophilic networks, questions around whether diasporic clusters are structural holes and the implications of these clusters being structural holes in the overall social networks of their 'host' and 'home' countries could meaningfully inform social policies that target these groups. We further explore triadic analysis in the subsection below.

Triadic analysis

Combinations of possible triads are significantly larger than combinations of dyads; a tie either exists or not in undirected dyads, and directed dyads have either one-way, two-way, or no tie. Triads are the subject of a foundational paper by Granovetter (1973) – now commonly referred to as the weak tie hypothesis. Granovetter (1973) defines a tie (and its strength) as, 'a combination of the amount of time, the emotional intensity, the intimacy (mutual confiding), and the reciprocal services which characterise the tie' (p. 1361). Using empirical evidence from a survey of job seekers, the paper suggests that there is a 'forbidden triad' where if A and B are connected, and A and C are connected, B and C will also be connected. This is particularly the case if the ties are strong between two people. Some ties can act as a 'bridge' (inhabiting a brokerage position) that spans parts of a social network and connect otherwise disconnected social groups; however, no strong tie can act as a bridge – only weak ties are brokers of relationships. The argument is that if someone is strongly tied to someone else, those around their tie will also be tied to them, and so these ties will be redundant. The weak tie hypothesis states that for the diffusion of information to happen across a network, we need weak ties rather than strong ties. The case study in this chapter explores the concept of weak ties in online diasporic traditional medicine discourse networks.

Triadic ties are also known in sociology as Simmelian ties. Georg Simmel observed that the enlargement of the smallest social group to a triad is a profound qualitative change of group condition that has ramifications for all three dyadic relationships and all three individuals (Simmel, 1922/1955). Triads cannot be understood as decomposable into three dyadic relationship systems; rather, they are themselves systems in which three dyadic systems are nested. As such, the triad is the basis of analyses that attend to structural forms. In a paper that empirically examines structural balance theory – which posits that a social network of interpersonal sentiments has a natural evolution towards particular generic forms of balanced social organisation at the macro level – Rawlings & Friedkin (2017) mobilised the triad as the steppingstone to understanding the macrostructure organisation of sentiment relations. Structural balance theory posits that some types of triads are forbidden while others are permitted based on four rules, with remarkable nonintuitive implications for the macrostructure of the group's sentiment network. Using the term 'friend' to designate a positive sentiment and the term 'enemy' to designate a negative sentiment, the classic balance model defines a sentiment network as balanced if it contains no violations of four assumptions (Rawlings & Friedkin, 2017):

- (A1) A friend of a friend is a friend,
- (A2) A friend of an enemy is an enemy,
- (A3) An enemy of a friend is an enemy,
- (A4) An enemy of an enemy is a friend.

Alternatively, these rules can be expressed in triad types in Table 2.1:

TABLE 2.1 Triad types in structural balance theory (adapted from Rawlings & Friedkin, 2017)

Triad type	Label	A4	A3	A2	A1
	300				
	102				
	003	Forbidden			
	120D			Forbidden	
	120U			Forbidden	
	030T	Forbidden	Forbidden	Forbidden	
	021D	Forbidden	Forbidden		
	021U	Forbidden	Forbidden	Forbidden	
	012	Forbidden			
	021C	Forbidden			Forbidden
	111U		Forbidden		Forbidden
	111D			Forbidden	Forbidden

(*Continued*)

TABLE 2.1 (Continued)

Triad type	Label	A4	A3	A2	A1
	030C		Forbidden		Forbidden
	201		Forbidden	Forbidden	Forbidden
	120C		Forbidden	Forbidden	Forbidden
	210		Forbidden	Forbidden	Forbidden

These rules are important to the construction of the concept of transitivity – another useful concept in social network analysis. (A1) is often referred to as the transitivity assumption; on the basis of this single rule, we can infer that (i) all macrostructures have one or more cliques (the entire structure may be one clique) within which relations are all positive; (ii) the relations between pairs of cliques are either all negative or asymmetric; (iii) every member of a clique has an identical sentiment towards every member of the macrostructure; and (iv) if cliques are joined by asymmetric relations, then a hierarchical form of macrostructure exists (Rawlings & Friedkin, 2017). The distinction is based on whether the configuration of sentiments in a triad contains a path involving all three members: $i \rightarrow j \rightarrow k$, or $j \rightarrow k \rightarrow i$, or $k \rightarrow i \rightarrow j$, and so on. If no such path exists in a triad, then the structural condition for a violation of transitivity does not exist, and the triad is defined as vacuously transitive because it does not violate transitivity (Rawlings & Friedkin, 2017). The probability of a positive $i \rightarrow k$ sentiment is more likely when a sequence $i \rightarrow j \rightarrow k$ of positive sentiments exists than when such a sequence does not exist, and the probability of an intransitive triad (one of the seven types of triads that violate transitivity) is lower than the probability of a triad that is not intransitive (one of the nine types of triads that do not violate transitivity).

Transitivity as a network concept has wide-ranging applications across different empirical domains. For researchers of diaspora, it is useful and important to understand how and why certain sentiments, attitudes, and behaviours spread across diasporic networks – a line of inquiry that can be productively engaged with through the lens of transitivity. From attitudes towards social change issues to voting behaviours, from political beliefs to generational trauma, researchers of diaspora can employ network transitivity as a robust and generative analytical tool for their particular research questions and contexts.

Meso-level analysis

Clusters – or communities – constitute the next level of network analysis. Clusters can be roughly understood as triads at scale. In network analysis, communities are understood in terms of modules and hierarchies. In this context, the term module is typically used to refer to a single cluster of nodes. Given a network that has been partitioned into non-overlapping modules in some fashion (although some methods also allow for over-lapping communities), we can continue dividing each module in an iterative fashion until each node is in its own singleton community (Porter et al., 2009). This partitioning process is hierarchical; at the end of this process, we are left with a hierarchy of nested modules. Community structure of a network refers to the set of graph partitions obtained at each 'reasonable' step of such procedures. Community detection can be applied individually to separate components of networks that are not connected (Porter et al., 2009).

In practice, clusters are identified with the help of community detection algorithms built into the most popular network analysis programs. Different algorithms would have different approaches to picking up clusters, but they all work off the notion of modularity – which is a metric that compares the number of edges within a cluster to the number of edges between clusters. Modularity can have a maximum score of 1, where all nodes are equally connected within and between clusters. A modularity of 0.3 is usually considered sufficient to detect communities (Hogan, 2017); the higher the modularity score, the more distinct the community. A score of 0 means that the communities are no more distinct than a random distribution of edges, and as such not picked up as a community. If most connections go between groups rather than within groups, then the modularity score is negative. Intuitively, a community is a cohesive group of nodes that are connected more densely to each other than to the nodes in other communities. The differences between many community detection methods ultimately come down to the precise definition of 'more densely' and the algorithmic heuristic followed to identify such sets (Porter et al., 2009).

There are many community detection algorithms, each comes with its own assumptions and operationalisations of network density. Here, I will introduce some well-established methods within the social sciences: k-means clustering (partitional clustering), centrality-based clustering, and the Leiden method. The simplest form of clustering is partitional clustering, which aims at partitioning a given data set into disjoint subsets so that specific clustering criteria are optimised. Given a set of observations (x_1, x_2, ..., x_n), where each observation is a d-dimensional real vector, k-means clustering aims to partition n observations into k ($\leq n$) sets $S = \{S_1, S_2, ...,S_k\}$ so as to minimise the within-cluster sum of squares. When working with a computer, a k-means algorithm is an iterative algorithm that partitions the dataset into k distinct non-overlapping subgroups where each data point belongs to only one group. k is a predefined value determined by separate estimation methods – an example of which will be featured in Chapter 3's case study.

Centrality-based clustering (also known as the Girvan-Newman algorithm) is a community detection algorithm based on betweenness centrality (Girvan & Newman, 2002). While we have previously covered the concept of high/low betweenness in nodes, an edge can also be said to have a high betweenness if it lies on a large number of paths between nodes. The betweenness of an edge quantifies traffic that goes through an edge as it connects one node to another by considering strictly shortest paths (geodesic betweenness) or densities of random walks (random walk betweenness) between each pair of nodes and averaging over all possible pairs. Communities can be detected through a process of ranking each of the edges based on their betweenness, removing the edge with the largest value, and recalculating the betweenness for the remaining edges. Recalculation is important as the removal of an edge can cause a previously low-traffic edge to have much higher traffic. An iterative implementation of these steps gives a divisive algorithm for detecting community structure, as it deconstructs the initial graph into progressively smaller connected chunks until one obtains a set of isolated nodes (Porter et al., 2009).

The Leiden method, developed by Traag et al. (2019), is intended as an improvement on the Louvain method – which was used in the case study below. In the Louvain method, the algorithm optimises a quality function such as modularity or Constant Potts Model (an improvement on modularity) in two phases: (1) local moving of nodes; and (2) aggregation of the network (Blondel et al., 2008). In the local moving phase, individual nodes are moved to the community that yields the largest increase in the quality function. In the aggregation phase, an aggregate network is created based on the partition obtained in the local moving phase. Each community in this partition becomes a node in the aggregate network. The two phases are repeated until the quality function cannot be increased further. The Leiden algorithm, by contrast, consists of three phases: (1) local moving of nodes, (2) refinement of the partition, and (3) aggregation of the network based on the refined partition, using the non-refined partition to create an initial partition for the aggregate network; as such, it is more complex than the Louvain algorithm (Traag et al., 2019). Both methods can be applied using the Gephi program or through publicly available Python and R programming language libraries.

Whole-network analysis

Whole-network analysis compares networks with other networks by drawing on meso-level concepts such as density and clustering, except applying them to the whole-of-network level. Density as a measure is best used when comparing networks of similar sizes or comparing the same network over time, as it can be misleading to compare the density of networks of substantially different sizes (Hogan, 2017). Global clustering coefficient – a scaled-up version of the previously discussed local clustering coefficient – allows us to measure the clustering in the whole network. The global clustering coefficient is the number of closed triads (triangles) over the total number of both open and closed triads; this measure can be applied to both undirected and directed networks. A large global clustering coefficient means that a network is highly clustered around a few nodes, and a low global clustering coefficient means that the edges that exist within a network are relatively evenly spread among all nodes.

There are also other ways to measure network clustering, including Watts and Strogatz (1998)'s average of the local clustering coefficients of all nodes n in a network, defined as:

$$\bar{C} = \frac{1}{n} \sum_{i=1}^{n} C_i$$

Compared to the global clustering coefficient measure, $\bar{C}$ places more weight on low-degree nodes while the global clustering coefficient places more weight on high-degree nodes. In practice, how we calculate the clustering coefficient depends on the context of the network we examine and the underlying phenomenon we want to quantify. Watts and Strogatz (1998)'s theorisation of the small-world network, defined as networks with a small average shortest path length and a large clustering coefficient, was partially based on the lived experience of real-world social networks where people who randomly run into each other at social events would end up talking about people they know in common (de Sola Pool & Kochen, 1978/2006). For early small-world network experiments, also see Travers and Milgram (1977). A more detailed discussion of small-world networks is included in Chapter 3.

Case study: weak ties and knowledge propagation across Vietnamese traditional medicine networks on Facebook

This section reports on the methods and summarises the key findings in Nguyen (2021a) as a case study for applying social network analysis to studying online diasporic discourse. In particular, this section introduces the domain-specific literature that the paper contributes to by briefly introducing its theoretical framework and its research questions before reporting in detail on the practical methodological choices made in its analysis. We also consider the results in light of how social network analysis works in tandem with other methods to reveal emergent socio-technical practices.

Case study overview

Social networking sites have emerged as popular platforms for decentralised exchange of information, including health information. Much of the literature on informational aspects of digital health focuses on a range of digital health-related resources for laypeople and opportunities for professional-to-professional communication. Online health information resources could provide a range of valuable social support – among which informational support is only one dimension (Coulson & Greenwood, 2012; Hether et al., 2016). Even weak ties that exist in a large network could bring about various forms of support, including access to different viewpoints, reduced communication risk, objective feedback from others, and reduced role obligations (Wright et al., 2010). Using social network analysis and content analysis, this paper examines the propagation of non-biomedical knowledge on Facebook – the largest social media platform in Vietnam. Vietnam is also Facebook's seventh-largest market (We Are Social, 2019). This case study provides a network map of user interaction on two Vietnamese traditional medicine social networking sites, as well as examining the patterns of support that the weakest ties – defined as participants who only engage in one interaction within their subnetworks – provide.

RQ1 What are the network characteristics of Vietnamese traditional medicine groups on Facebook?

RQ2 What social support is being exchanged among weak ties on Vietnamese traditional medicine groups on Facebook?

Sample selection and data collection

This research employs a multiple case study approach by examining two public Facebook sites that focus on Vietnamese traditional medicine. A sampling frame of 1900 Facebook sites that focused on Vietnamese traditional medicine was identified through a keyword search using Facebook's built-in search engine. Of these sites, two public Vietnamese traditional medicine sites were selected for analysis based on their popularity (measured in number of active members), level of member engagement (measured in number of posts per week), and privacy nature (closed, secret, or public site). These sites are numbered Site 1 and Site 2, respectively. Both sites are public sites with over 30,000 active members and a robust interaction frequency of approximately 300 posts per month. The sample included a total of 6,923 original posts and 23,333 associated comments, collected from the two sites over 3 years (with a cut-off date of 25 March 2019). The cut-off date is the same for both sites, whose start dates are within months of each other so that data collection can provide a complete snapshot of each page's trajectory.

The body of data gathered in the 3 years up until the cut-off date was considered sufficiently robust for meaningful analysis. Data were collected with a purpose-built web scraper. The scraper utilises the Puppeteer library developed by Google to collect publicly available data via the Chrome web browser. The scraper collected all text content on original posts and their associated comments. Social network analysis was performed on this sample. Following this quantitative analysis, a sub-sample of 2,570 comments and 1,543 associated original posts was retrieved for content analysis. This sub-sample contains comments that are made by nodes with a degree of one – which represents Facebook users who only make one comment on either an original post or a post comment across four subnetworks. These nodes represent the weakest ties in their respective subnetworks.

Application of social network analysis

Social network data were collected from the main sample. A list of 386 keywords was created using the appendix from Circular 05/2015/TT-BYT which includes comprehensive variations and specifications of officially recognised traditional medicines (for the complete list, refer to Nguyen, 2021a). This keyword list includes more granular keywords that capture the nuances of everyday use in Vietnam. For example, hà thủ ô đỏ (tuber fleeceflower) and hà thủ ô đỏ chế (processed tuber fleeceflower) were considered two separate keywords. Posts and associated comments that return at least one keyword match are collected to represent discourse that exists in accordance with government regulation of traditional medicine (subset 1). The rest of the sample with no keyword match is labelled subset 2, which represents discourse that exists beyond government purview. Given the sample size of 30,000+ units of analysis, this quantitative data scan helps establish preliminary parameters around the nature of discourse on these sites.

Network structure was assessed with network size (number of nodes and edges in the network), average path length (the average number of steps along the shortest paths for all possible pairs of network nodes), and betweenness centrality (the number of times a node acts as a bridge along the shortest path between two other nodes). Network modularity was calculated using the Louvain method (Blondel et al., 2008). Modularity is a measure used to quantify the quality of a given division of a network into different communities (Li & Schurmann, 2011). The Louvain algorithm is one of the fastest and most effective modularity-based algorithms to detect communities in large networks (Orman et al., 2011). Communities, or clusters, are interesting in the study of complex networks because it gives us an indication of the network's structural cohesion.

Content analysis methodology

A content analysis coding scheme and a coding manual were developed based on Hether et al. (2016, p. 5)'s coding scheme for support-seeking messages on social networking sites. Hether et al. (2016)'s coding scheme was developed based on the Optimal Matching Model literature. Five dimensions of social support were coded: informational; emotional; esteem; network; and tangible. Support-seeking and support-provision are also coded separately. Please refer to Table 2.2 for the social support coding scheme for support-seeking comments and posts with examples from the dataset. A similar scheme, though not included here, was also used for support-provision posts and comments.

These instruments were pilot-tested through two rounds of consensus coding, where two research assistants independently coded a subsample of posts and comments before meeting to discuss the pilot results. Discrepancies were identified, refined, and resolved by comparing how each coder interpreted and applied the coding scheme. Every effort was made to be consistent with earlier research and retain the validity of the social support taxonomy.

Interrater reliability was assessed on an independent sample of 10% of the posts after pilot testing. Two coders shared the same tertiary education background and were bilingual in English and Vietnamese. Reliability was computed with simple percentage agreement and Cohen's kappa κ (Cohen, 1960, 1968). While simple percentage agreement is directly interpretable, it does not account for the possibility of agreement occurring by chance and is therefore considered too liberal as a measure of interrater reliability (Lombard, 2002). Cohen's kappa κ is a robust interrater and intrarater reliability statistic that accounts for chance agreement and is widely used in social sciences (Neuendorf, 2017). Both indices are reported here. For the seeking and providing of each dimension of support, reliability was as follows: (1) informational support, 100% and κ = 1.00; (2) emotional support, 95.24% and κ = 0.972 (p = 0); (3) esteem support, 100% and κ = 1.00; (4) relationship support, 100% and κ = 1.00; and (5) tangible support, 100% and κ = 1.00 in each reliability sample.

Social network analysis results

1 Network scan results

The automated keyword scan divides the sample into two subsets (see Figure 2.2). Subset 1 contains 1,299 posts and 12,240 associated comments that have at least one keyword match. This subset is further divided into sub-network A (matches within site 1) and sub-network C (matches within site 2). Subset 2 contains 5,624 posts and 12,200 comments with no keyword match. This subset is further divided into sub-network B (non-matches within site 1) and sub-network D (non-matches within site 2). These results show that a slight majority of information being exchanged on traditional medicine sites does not include the current language used by government regulation. Of the 386 keywords used to scan the sample, Site 1 returns 197 matches

TABLE 2.2 Coding scheme for social support dimensions, adapted from Hether et al. (2016)

Dimension	Description	Examples
Informational support	Requests for knowledge or information about a specific issue or situation.	• Suggestions/advice: Someone to offer ideas and suggest actions (e.g., "How should I incorporate traditional medicine into my Western medical regimen?"). • Referrals: Referral to another source of information, such as a website or book (but excluding a person/network connection) (e.g., "Where can I find family recipes for centella?"). • Situation appraisal: Someone to reassess or redefine the situation (e.g., "Isn't Okinawan spinach supposed to be softer-looking?"). • Teaching: Someone to explain the facts, or news about a situation or about the skills needed to deal with the situation (e.g., "What can Chinese mugwort cure?").
Emotional support	Problem is framed in terms of emotion and participant is seeking some kind of emotional feedback or understanding. The focus of the message will typically be on feelings, not information.	• Affection: Includes "virtual" physical contact such as sending hugs, using emoticons and Facebook 'stickers'. Generally, this kind of support is provided, without being asked for (e.g., ☺). • Sympathy: Someone to feel sorrow or regret for the support-seeker's situation (e.g., "I have hemorrhoids and because it's a sensitive condition I am embarrassed to talk about it"). • Understanding/empathy: Someone to understand the situation, often through personal experience (e.g., "Does anyone here also have these small dots on their arms?!!"). • Encouragement: Someone to provide the recipient with hope and confidence (e.g., "Wish me luck so I could keep trying to control this condition"). • Prayers: Someone to pray with/for another member (e.g., "*Nam Myōhō Renge Kyō*, please pray for me so I could overcome my sickness").
Esteem support	Refers to regard for one's skills, abilities, and intrinsic value. It is distinguished from emotional support because it focuses on a person's self-perceptions rather than their emotions about something else.	• Validation: Someone to express agreement with the support-seeker's perspective on the situation (e.g., "I've been using cassia grandis plant to cure constipation, is this the correct way?"). • Relief of blame: Someone to alleviate the support-seeker's feelings of guilt about the situation (e.g., "I've been eating so much spicy food lately, this has caused an acne outbreak on my face. Does anyone also eat a lot of spicy food? Is it bad for you?").

(Continued)

TABLE 2.2 (Continued)

Dimension	Description	Examples
		• Compliment (coded in the provision of support only): Saying positive things about a support-seeker (e.g., "That's the right way to stop diarrhea, just keep using the plant for a few more days").
Network support	Attempts to create structural connections with other individuals or groups	• Access: Someone to provide access to new companions, including access to another support group (e.g., "Can someone please point me to authentic soapnut sellers?").
		• Friendship: An explicit request to make friends and participate in the group. Also includes any exchange of personal email addresses or requests/offers to talk at a later date or in a private conversation (e.g., "I just joined this group, thanks admins for accepting my request. I look forward to knowing all of you').
		• Physical presence: Someone looking to meet other women in the same geographic location (e.g., "I'm living in Can Tho and looking for a good doctor of eastern medicine").
		• Companions: Someone who needs to be reminded of the availability of existing companions with whom they can talk to about a problem (e.g., "I have this problem but I don't know how to share it. Could someone talk to me?).
Tangible support	Requests for physical aid	• Loan: A request to lend the support-seeker something including money. (e.g., Does anyone have *dien chan* tools that I can borrow for two weeks? I live in Rach Gia")
		• Gift: A request to give the support-seeker something, including money. (e.g., "Does someone have lotus roots that they can spare? It's for urgent use")
		• Direct task: A request to perform a task directly for the support-seeker (e.g., "I want to buy some false ginseng for my stomachache, please send me a direct message if you're selling").

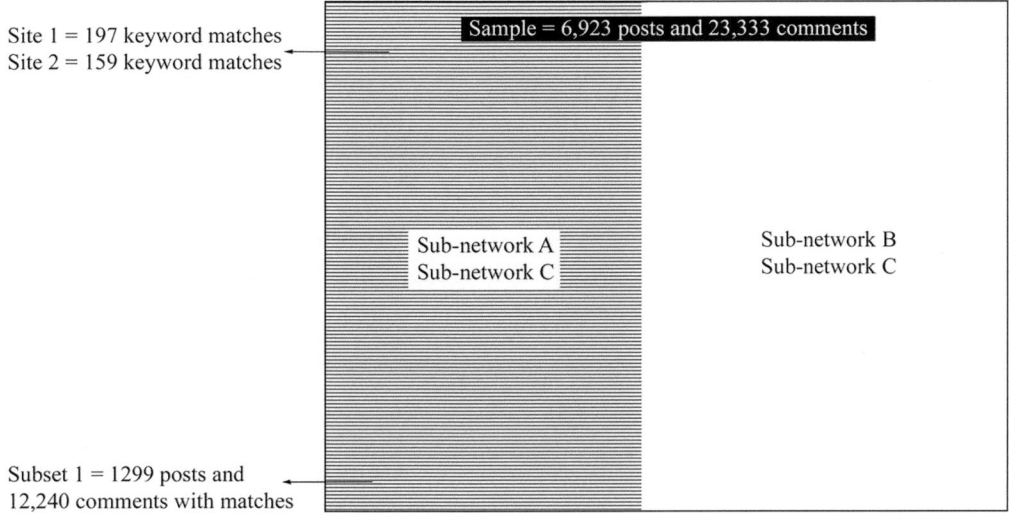

Site 1 = 197 keyword matches
Site 2 = 159 keyword matches

Sample = 6,923 posts and 23,333 comments

Sub-network A
Sub-network C

Sub-network B
Sub-network C

Subset 1 = 1299 posts and
12,240 comments with matches

FIGURE 2.2 Network scan results.

while Site 2 returns 159 matches. These matches represent roughly half of the official terminologies used by the Vietnamese government to regulate traditional medicine.

From these results, we could see that the current language used in regulation does not sufficiently overlap with the language used by everyday Vietnamese. This could partly be explained by the formal (usually Sino-Vietnamese) language used in Vietnamese legal documents, which differs from the highly variable everyday terms used for the same or similar medical ingredients across different geographical areas in Vietnam. Twenty keywords with the highest number of matches in original posts and associated comments across both sites are reported in Table 2.3, together with their regulated treatment and recognised effects.

While six of the top twenty herbal plants have recognised effects that are congruent with the biomedical philosophy of the body (laxative, hemostasis, anti-inflammatory, diuretics, aphrodisiac), the medicinal property of a majority of these plants can only be understood in the context of local cosmos. Concepts such as 'heat', 'wind', 'energy', and 'blood' have very specific meanings in Vietnamese traditional medicine, with influences from Chinese traditional medicine (Monnais et al., 2012). 'Cold wind', for example, is believed to cause fever, headaches, and general aches, while 'damp wind' is believed to be the reason for fatigue, nausea, or diarrhea (Dashtdar et al., 2016). 'Blood deficiency' also carries highly specific meanings; according to Vietnamese traditional medicine, similar to Chinese traditional medicine, failure to 'nourish' blood might lead to 'blood stasis', which is believed to cause limb stiffness and pain (Dashtdar et al., 2016). These contextually contingent narratives of health and illness seem to dominate the discourse being exchanged in the current sample. This suggests that, instead of simply offering commensurable traditional equivalents to their biomedical counterparts, medical knowledge of the kind being propagated among the online groups studied here also carries with them worldviews and beliefs that have been displaced by the scientific biomedical enterprise. The online propagation of these knowledges could translate to a new techno-social life of marginalised systems of cultural belief, which warrants further investigation in future studies.

TABLE 2.3 Top twenty keyword matches

Keyword	Original post count	Comment count	Traditional medicine treatment/effect	Group number *
"Mật ong" (Honey)	67	188	Laxative	XXI
"Cam thảo" (Chinese liquorice)	42	55	Vital energy	XXVIII
			Clears heat and removes toxicity	VII
"Ngải cứu" (Mugwort)	32	42	Hemostasis	XVIII
"Tam thất" (Panax pseudoginseng)	26	46	Hemostasis	XVIII
"Dâu" (Mulberry)	25	63	Nourishing blood	XXV
"Đinh lăng" (Ming aralia)	22	43	Vital energy	XXVIII
"Lá lốt" (Piper lolot)	22	48	Anti-inflammatory	III
"Đậu đen" (Catjang)	19	34	Dispels wind-heat	II
"Xạ đen" (Celastrus hindsii)	17	37	Clears heat and removes toxicity	VII
"Diếp cá" (Fish mint)	16	35	Clears heat and removes toxicity	VII
"Mã đề" (Broadleaf plantain)	16	31	Diuretics	XIX
"Trinh nữ" (Shameplant)	15	34	Anti-inflammatory	III
"Ba kích" (Indian mulberry)	12	17	Aphrodisiac	XXVII
"Bồ kết" (Honey locust)	11	29	Regain consciousness	XV
"Cỏ xước" (Chaff-flower)	10	16	Activates blood	XVII
"Cà gai leo" (Solanum)	8	21	Anti-inflammatory	III
"Rau má" (Centella)	8	23	Clears heat and evil wind	IX
"Trầu không" (Betel)	6	30	Dispels wind-cold	I
"Tràm" (Weeping paperbark)	5	22	Dispels wind-cold	I
"Cải trời" (Bitter lettuce)	5	21	Clears heat and dispels wind-heat	VIII

Notes
* *according to Circular 05/2015/TT-BYT classification.*

2 Formal network measures

Four subnetworks are identified as a result of the keyword scan, labelled A – D. Table 2.4 summarises the formal network measures assessed.

Despite accounting for roughly 46% of all posts within the sample, subnetworks A and C are more densely populated than subnetworks B and D – with significantly higher numbers of nodes and edges. Within these subnetworks, there are not only more network interactions but also more actors involved in these interactions. The average path lengths between two nodes in subnetworks A and C are also shorter than those in subnetworks B

TABLE 2.4 Report of global network measures

Subnetwork	Number of node	Number of edge	Average path length	Network modularity	Number of community
A	3785	7722	4.44	0.59	21
B	1657	1739	7.39	0.89	142
C	2111	3086	5.13	0.71	32
D	1793	2132	7.53	0.83	41

and D – this means that information within these subnetworks reaches more people in a shorter time period. Network detection using the Louvain method with a resolution of 1.0 across all four subnetworks shows that more communities are found within subnetworks B and D; this is particularly pronounced between A and B, where the number of communities detected in B is more than six times that detected in A. Subnetworks B and D also have higher modularity; this means that the connections within these subnetwork communities are much denser than those outside of these communities. Communities found within subnetworks B and D are harder to break than those found in subnetworks A and C.

A visualisation of these subnetworks is provided in Figure 2.3. Due to the high number of communities detected, only the ten largest communities are represented in colour. The size of each node represents betweenness centrality. Fan-like subgraphs such as those boxed in red in Figure 2.3 represent nodes that act as a hub of other nodes with a degree of one. Nodes with a degree of one are connected to only one other node within the network. In the current context, nodes with a degree of one represent Facebook users who only ever respond to one original post or post comment and who, other than that interaction, have no other engagement in their subnetworks. These 'transient users' are well-recognised in literature as a key characteristic of Facebook sociality. These nodes can be said to have the weakest ties to their subnetworks; they also usually interact with the same nodes that have the highest betweenness centrality. Subnetworks B and D have a higher percentage of nodes and edges of this nature. Specifically, around 20% of subnetworks B and C are made up of single-degree nodes. Only 0.3% and around 3% of subnetworks A and D are single-degree nodes, respectively.

The prevalence of weak ties in subnetworks B and D warrants further examination. Within these subnetworks, no mention of any keywords was found. These subnetworks also have a high number of dispersed communities detected; this means that individuals engaging in social support in these subnetworks organise themselves in multiple, much less rigorous discourse communities. Previous literature, however, has indicated that weak ties can provide significant benefits to their networks. The next section reports on the social support provided by weak ties in these subnetworks.

Content analysis results

Overall, support provision (63.2%) is more prevalent than support-seeking activity among weak ties. Informational support is the most prominent form of support sought and provided across two sampled sites and among all four subnetworks, followed by emotional support, tangible support, network support, and esteem support (see Figure 2.4). This

A.   B.

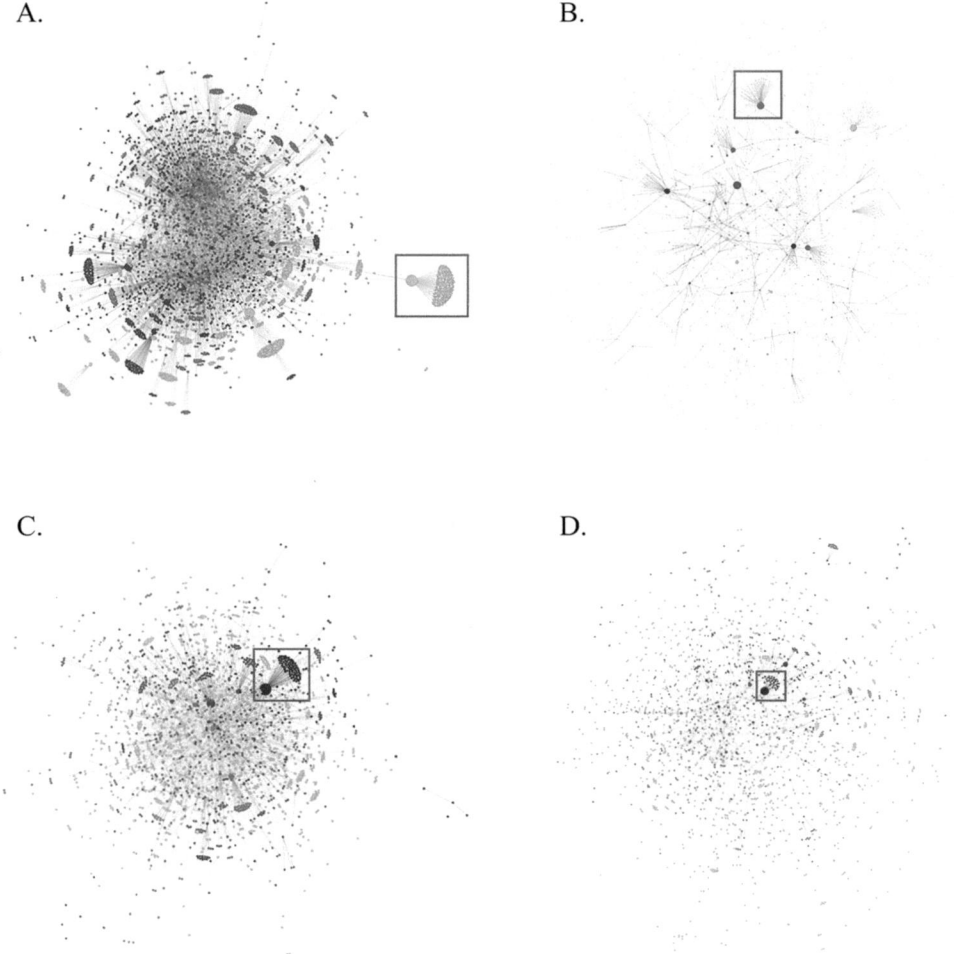

C.   D.

FIGURE 2.3 Social network graphs of four sub-networks across two sites.

overall ranking is also consistent with previous research into social support on health-related social networking sites (see Discussion). Within informational support, teaching and suggestion/advice constitute the overwhelming majority of all support seeking and provision across all subnetworks, followed by referrals and situation appraisal. Affection and empathy are the most prominent types of emotional support, while gifting and direct tasks are the most common tangible support provided and sought among weak ties in the sampled sites. Network support involves only two dimensions, namely access and physical presence support, while esteem support involves validation and compliments.

Implications of research findings

Within subnetworks whose discourses lie outside of the current Vietnamese regulatory frameworks for traditional medicine, more and stronger communities of discourse are found compared to subnetworks whose discourses utilise the language used in current

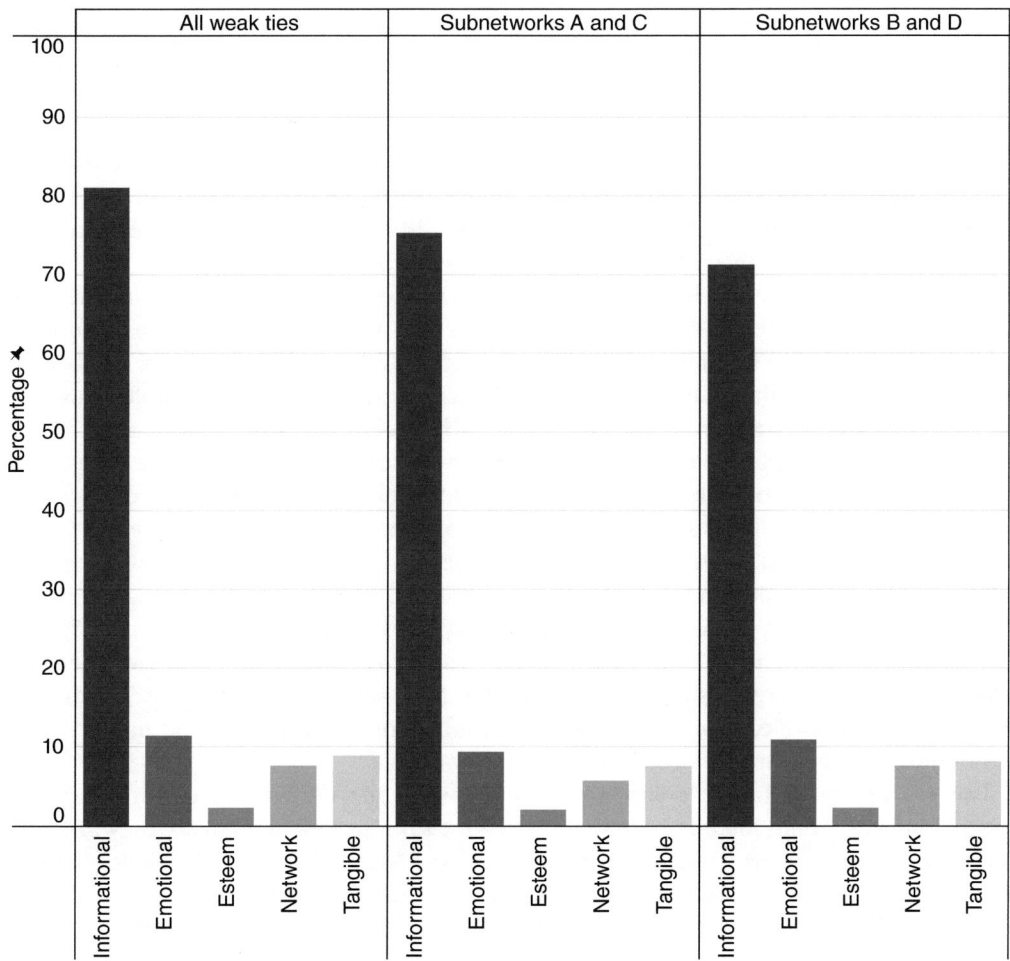

FIGURE 2.4 Social support types among weak ties.

Vietnamese regulations. A significantly higher number of weak ties are also found in these subnetworks; these weak ties provide a range of social support to users within the sites sampled – most dominantly information support. This could be a cause for concern, considering how the veracity of the informational support provided is not currently recognised or regulated on a governmental level. Given that nurturant forms of support are much less prevalent, the impact that online support and knowledge-sharing groups have on their participants could be well beyond that of a moral nature. The prevalence of weak ties among discourses outside of existing regulatory frameworks can be indicative of a network dynamic that corresponds with the double stigma of firstly seeking and giving online support for health conditions in non-biomedical knowledge settings, and secondly of participating in non-biomedical knowledge discourses that are not officially recognised. Both stigmas are well documented in the context of Vietnam; while the documentation and regulation of traditional medicine in Vietnam is substantial, the prevalence of unregulated folk medicine remains extensive (Thompson, 2017). This further highlights the complex coexistence of biomedical and non-biomedical modalities as the latter continue to evolve and carve out alternative discourse spaces for their survival. Weak ties have minimal communication risks and role obligations while being perceived as offering unique viewpoints that are more objective (Wright, 2016). All these peculiar dynamics could lead to a mechanism where participants with little investment in maintaining sustained contribution to network discourse propagate unverified (and unverifiable) information with little burden for potential consequences.

While the prevalence of informational support found here is consistent with previous research into other health-related social networking sites such as psychosis support groups (Chang, 2009) or prenatal sites (Hether et al., 2016), the prevalence of tangible support is a novel finding. Chang (2009) and Hether et al. (2016), for example, found no evidence of tangible support sought or provided in their studies in Taiwan and the US, while tangible support is a significant dimension of support in Vietnamese traditional medicine sites. This could be explained by the long history of traditional medicine in Vietnam as an informal and sometimes philanthropic healthcare activity, wherein people volunteer raw medical ingredients that they have in their possession as a way to help others. By the same measure, people who are in need of these materials also seek out those who grow, hunt, or collect these medicinal plants to obtain them either for free or for a price. These activities seem to be emigrating online, where the same offline, community-bound activities are being facilitated over social networking sites; 'communities' that exist online, however, have a much larger reach and scale, and might not be as bounded by geographical and sociocultural borders.

The prevalence of tangible support, together with the dominance of informational support on traditional medicine sites, also account for a unique dynamic – they are both action-facilitating forms of support, which tend to foster behaviours that help mitigate health stressors rather than coping with the consequences of these stressors. While the overall ranking of each social support dimension is consistent with what previous studies have found (Coulson & Greenwood, 2012; Gray, 2013; Loane & D'Alessandro, 2013; Hether et al., 2016), the proportion of emotional support compared to tangible support found here is much lower than in previous studies. This means that there seems to be more significant tangible support within traditional medicine networks in Vietnam compared to other health-related networks on social media. Particularly, the tangible support sought and volunteered with the sample comes mostly in the form of gift-giving and gift-seeking (asking for/giving out raw herbs and raw medicinal plants). These acts of tangible support are common in the Vietnamese context outside of and before social media at the neighbourhood and village level; that they are now happening via the mediation of social media, which is less bounded by geography, could signal a significant change in the way support communities are being formed. This warrants further research into whether any change in scale facilitated by social media also leads to potential changes in the quantity and quality of health support networks in Vietnam. Considering these findings against the context of Vietnam, where the healthcare system is becoming less and less affordable as the government progressively removes state subsidy for healthcare services – coupled with rising out-of-pocket healthcare expenditure – it seems to be the case that Vietnamese people are turning to social media to seek alternative courses of action in order to mitigate these trends. This turn to social media should be understood as a continuation of existing cultural practices concerning the management of health and illness among Vietnamese which predate social media, rather than a new set of practices brought about by digital technologies.

The dominance of contextually contingent beliefs about health and illness in Vietnam, inferred from the most frequently occurring names of herbal plants in the dataset, is also an important finding. This is pointing to the persistence of culturally grounded beliefs and knowledge, which have found digital expressions through social media. Although this study looked specifically at textual expressions of this knowledge, multimedia forms of expression such as images, prerecorded videos, live streaming videos, and synchronous watching of pre-recorded video also constitute an important part of this digital knowledge ecology. Future research should turn their attention to how changing forms and expressions of knowledge could potentially alter its substance and sociality. While limited by the sites sampled, the analysis presented here provides an empirically driven account of online propagation of traditional knowledge and the types of support they facilitate. The goal was not to exhaust all possible social media content, but rather to provide rigorous analysis and suggest future directions in an under-researched topic that could have important implications in different disciplines. Future research into networks of knowledge propagation on social media could further mobilise digital methods such as natural language processing techniques to study the topics of discourse among these online communities in a more systematic fashion.

Conclusion

This chapter has provided a discussion on the basics of social network analysis as a method in relation to the study of digital diaspora, highlighted the deliberative nature of this method in setting parameters around constructed networks, outlined some implications that this method could have for the general program of diaspora studies, and presented a case study of social network analysis as applied to an online network of traditional medicine discourse among mainland and diasporic Vietnamese communities. In the next chapter, we will consider linguistic issues facing researchers of diaspora in conjunction with a fast-growing area of research at the intersection of linguistics and computer science – that of natural language processing.

References

Blex, C., & Yasseri, T. (2022). Positive algorithmic bias cannot stop fragmentation in homophilic networks. *The Journal of Mathematical Sociology*, *46*(1), 80–97. 10.1080/0022250X.2020. 1818078.

Blondel, V. D., Guillaume, J. L., Lambiotte, R., & Lefebvre, E. (2008). Fast unfolding of communities in large networks. *Journal of Statistical Mechanics: Theory and Experiment*, *2008*(10), P10008. 10.1088/1742-5468/2008/10/P10008.

Burt, R. S. (2000). The network structure of social capital. *Research in organizational behavior*, *22*, 345–423. 10.1016/S0191-3085(00)22009-1.

Chang, H. J. (2009). Online supportive interactions: Using a network approach to examine communication patterns within a psychosis social support group in Taiwan. *Journal of the American Society for Information Science and Technology*, *60*(7), 1504–1517. 10.1002/ asi.21070.

Cohen, J. (1960). A coefficient of agreement for nominal scales. *Educational and Psychological Measurement*, 20(1), 37–46.

Cohen, J. (1968). Weighted kappa: Nominal scale agreement with provision for scaled disagreement or partial credit. *Psychological Bulletin*, 70(4), 213–220. 10.1037/h0026256.

Coulson, N. S., & Greenwood, N. (2012). Families affected by childhood cancer: An analysis of the provision of social support within online support groups. *Child: Care, Health and Development*, 38(6), 870–877. 10.1111/j.1365-2214.2011.01316.x.

Dashtdar, M., Dashtdar, M. R., Dashtdar, B., Kardi, K., & Khabaz Shirazi, M. (2016). The concept of wind in traditional Chinese medicine. *Journal of Pharmacopuncture*, 19(4), 293–302. 10.3831/KPI.2016.19.030.

de Sola Pool, I., & Kochen, M. (2006). Contacts and influence. In M Newman, A-L Barabási, & D J Watts (Eds.), *The structure and dynamics of networks* (pp. 83–129). Princeton University Press. (Reprinted from *Social Network*, 1(1), pp. 5–15, 1978).

Dinh, R., Gildersleve, P., Blex, C., & Yasseri, T. (2022). Computational courtship understanding the evolution of online dating through large-scale data analysis. *Journal of Computational Social Science*, 5(1), 401–426. 10.1007/s42001-021-00132-w.

Elliott, M., & Golub, B. (2013). A network approach to public goods. In *EC '13 Proceedings of the fourteenth ACM conference on Electronic commerce* (pp. 377–378). ACM. https://resolver.caltech.edu/CaltechAUTHORS:20131008-153813901.

Falk, A., & Fischbacher, U. (2006). A theory of reciprocity. *Games and Economic Behavior*, 54(2), 293–315. 10.1016/j.geb.2005.03.001.

Freeman, L. C. (1978/79). Centrality in social networks: conceptual clarification. *Social Networks*, 1(3), 215–239. 10.1016/0378-8733(78)90021-7.

Girvan, M., & Newman, M. E. (2002). Community structure in social and biological networks. *Proceedings of the National Academy of Sciences*, 99(12), 7821–7826. 10.1073/pnas.122653799.

Granovetter, M. S. (1973). The strength of weak ties. *American Journal of Sociology*, 78(6), 1360–1380.

Gray, J. (2013). Feeding On the web: Online social support in the breastfeeding context. *Communication Research Reports*, 30(1), 1–11. 10.1080/08824096.2012.746219.

Hennig, M., Brandes, U., Pfeffer, J., & Mergel, I. (2013). *Studying social networks: A guide to empirical research*. Campus Verlag.

Hether, H. J., Murphy, S. T., & Valente, T. W. (2016). A social network analysis of supportive interactions on prenatal sites. *Digital Health*, 2, 1–12. 10.1177/2055207616628700.

Hogan, B. (2017). Online social networks: Concepts for data collection and analysis. In N G Fielding, R Lee, & G. Blank (Eds), *The sage handbook of online research methods, second edition* (pp. 241–258). Sage Publications, Thousand Oaks, CA.

Holland, P. W., & Leinhardt, S. (1971). Transitivity in structural models of small groups. *Comparative Group Studies*, 2(2), 107–124. 10.1177/104649647100200201.

Kruja, E., Marks, J., Blair, A., & Waters, R. (2002). A short note on the history of graph drawing. In P Mutzel, M Jünger, & S Leipert (Eds.), *Graph Drawing: 9th International Symposium, GD 2001 Vienna, Austria, September 23-26, 2001, Revised Papers* (Vol. 2265) (pp. 272–286). Springer.

Labun, A., & Wittek, R. (2014). Structural holes. In R. Alhajj, & J Rokne (Eds.), *Encyclopedia of social network analysis and mining*. Springer.

Laniado, D., Volkovich, Y., Kappler, K., & Kaltenbrunner, A. (2016). Gender homophily in online dyadic and triadic relationships. *EPJ Data Science*, 5(1), 19. 10.1140/epjds/s13688-016-0080-6.

Li, W., & Schurmann, D. (2011). Modular community detection in networks. In T Walsh (Ed.), *Proceedings of the Twenty-Second International Joint Conference on Artificial Intelligence* (pp. 1366–1371). AAAI Press.

Loane, S. S., & D'alessandro, S. (2013). Communication that changes lives: Social support within an online health community for ALS. *Communication Quarterly*, *61*(2), 236–251. 10.1080/01463373.2012.752397.

Lombard, M., Snyder-Duch, J., & Bracken, C. C. (2002). Content analysis in mass communication: Assessment and reporting of intercoder reliability. *Human Communication Research*, *28*(4), 587–604.

McPherson, M., Smith-Lovin, L., & Cook, J. M. (2001). Birds of a feather: Homophily in social networks. *Annual Review of Sociology*, 415–444. 10.1146/annurev.soc.27.1.415.

Molm, L. D., Collett, J. L., & Schaefer, D. R. (2007). Building solidarity through generalized exchange: A theory of reciprocity. *American Journal of Sociology*, *113*(1), 205–242. 10.1086/517900.

Monnais, L., Thompson, C. M., & Wahlberg, A. (Eds.). (2012). *Southern medicine for southern people: Vietnamese medicine in the making*. Cambridge Scholars Publishing.

Neuendorf, K. A. (2017). *The content analysis guidebook*. SAGE.

Newman, M. E. (2005). A measure of betweenness centrality based on random walks. *Social Networks*, *27*(1), 39–54. 10.1016/j.socnet.2004.11.009.

Nguyen, D. (2021a). Dropping in, helping out: Social support and weak ties on traditional medicine social networking sites. *Howard Journal of Communications*, *32*(3), 235–252. 10.1080/10646175.2021.1878478.

Orman, G. K., Labatut, V., & Cherifi, H. (2011). Qualitative comparison of community detection algorithms. In *International conference on digital information and communication technology and its applications* (pp. 265–279). Springer.

Park, S. H., & Luo, Y. (2001). Guanxi and organizational dynamics: Organizational networking in Chinese firms. *Strategic Management Journal*, *22*(5), 455–477. 10.1002/smj.167.

Porter, M. A., Onnela, J. P., & Mucha, P. J. (2009). Communities in networks. *Notices of the AMS*, *56*(9), 1082–1097.

Rawlings, C. M., & Friedkin, N. E. (2017). The structural balance theory of sentiment networks: Elaboration and test. *American Journal of Sociology*, *123*(2), 510–548. 10.1086/692757.

Simmel, G. (1955). *Conflict and the web of group affiliations* (R Bendix, Trans.). Free Press. (Original work published 1922).

Starr Jr, R. G., Zhu, A. Q., Frethey-Bentham, C., & Brodie, R. J. (2020). Peer-to-peer interactions in the sharing economy: Exploring the role of reciprocity within a Chinese social network. *Australasian Marketing Journal (AMJ)*, *28*(3), 67–80.

Surma, J. (2016). Social exchange in online social networks. The reciprocity phenomenon on Facebook. *Computer Communications*, *73*, 342–346. 10.1016/j.comcom.2015.06.017.

Thompson, C. M. (2017). The implications of gia truyền: Family transmission texts, medical authors, and social class within the healing community in Vietnam. *South East Asia Research*, *25*(1), 34–46. 10.1177/0967828X17690045.

Traag, V. A., Waltman, L., & Van Eck, N. J. (2019). From Louvain to Leiden: guaranteeing well-connected communities. *Scientific Reports*, *9*(1), 1–12.

Travers, J., & Milgram, S. (1977). An experimental study of the small world problem. In S Leinhardt (Ed.), *Social networks: a developing paradigm* (pp. 179–197). Academic Press.

Wagner, D. (2003). Lecture 6: An introduction to graph. CS70: Discrete Mathematics for CS. https://people.eecs.berkeley.edu/~daw/teaching/cs70-f03/Notes/hypercube.pdf.

Watts, D. J., & Strogatz, S. H. (1998). Collective dynamics of 'small world' networks. *Nature*, *393*(6684), 440–442.

We Are Social. (2019). *Digital 2019: Q3 global digital statshot*. DataReportal. https://datareportal.com/reports/digital-2019-q3-global-digital-statshot.

Wright, K. (2016). Social networks, interpersonal social support, and health outcomes: A health communication perspective. *Frontiers in Communication*, *1*, 10. 10.3389/fcomm.2016.00010.

Wright, K. B., Rains, S., & Banas, J. (2010). Weak-tie support network preference and perceived life stress among participants in health-related, computer-mediated support groups. *Journal of Computer-Mediated Communication*, *15*(4), 606–624.

3

NATURAL LANGUAGE PROCESSING AND LINGUISTIC ISSUES IN DIASPORA RESEARCH

Introduction

Natural language processing (NLP) – also known as computational linguistics – is a field of study that involves the engineering of computational models and processes to solve practical problems in understanding human languages (Otter et al., 2021). NLP has a wide variety of commercial applications, which drive the automated machines we encounter every day: machine translation (e.g., Google Translate), speech technologies (e.g., voice assistant technologies such as Siri and Alexa), dialogue interfaces (e.g., chatbots on websites), text analytics (e.g., topic modelling of social media data), natural language generation (e.g., automated image captioning, automated data-driven news stories), and writing assistance (e.g., spell checkers, grammar checkers). NLP research is a burgeoning area of field with a rich and dynamic history, going back to early works on machine translation in the late 1940s. While early NLP research focused on machine translation (1940s–1960s), developments in this area went through radical changes with the influence of artificial intelligence from 1960s–1970s, the adoption of logico-grammatical style NLP from 1970s–1980s, before the 'massive data-bashing period' began from the 1980s onwards (Jones, 1994). NLP as we know it today invests in machine learning and deep learning approaches that allow us to handle large amounts of complex, multimodal texts.

While the 'conspicuous move into statistical language data processing' in the 1980s – as Jones (1994) put it – started with a rapid growth in the supply of machine-readable structured texts in service of information management, this approach became increasingly ingrained in the way NLP research is currently organised as NLP is tasked with extracting useful information from multimodal, multimedia texts that are unstructured in nature – such as social media data. Since the 1980s, NLP increasingly relies on data-driven computation involving statistics, probability, and machine learning (Otter et al., 2021). Deep learning – which uses artificial neural networks to enable billions of trainable parameters for text processing – has been made possible in NLP research thanks to increases in computational power in graphical processing units (GPUs). Neural models represent a step

DOI: 10.4324/9781003336556-5

change in the way NLP research and applications are heading; hardware made specifically for training deep networks – such as Google's Tensor Processing Unit (TPU), Microsoft's Catapult, and Intel's Lake Crest – are being made more widely available both for research and commercial purposes. This 'neural turn' has also ushered in the resurgence of neuromorphic computing, which implements neural structures at the hardware level with the development of neuromorphic chips (Otter et al., 2021).

For researchers of diasporas, NLP presents exciting opportunities to explore large corpora of text produced by diasporas across different languages in an automated fashion. This is particularly pertinent in the context of social media data produced by diaspora community groups, where researchers are faced with vast amounts of digital text as traces left behind by diasporic conversations and discussions. NLP techniques offer ways to process these texts in formal and robust ways; however, it should be noted that while the processing of these texts can be automated with NLP models, their interpretation cannot. NLP assists diaspora researchers by allowing them to systematically organise large volumes of text into useful categories that can then be variously interpreted according to their research questions, in specific contexts. While NLP cannot replace established practices of qualitative coding, it nevertheless offers researchers new and useful tools to organise and infer meaning from their data. Researchers of diaspora should approach NLP as one way with which they can tackle the difficult task of text analysis within their overall research toolkit – and as an approach that allows them to ask slightly different questions, as well as to answer familiar questions slightly differently.

NLP and social media data: the case for digital diaspora research

The proliferation of social media data – and more generally user-generated data – has posed as many interesting challenges as opportunities within NLP research. On one hand, the production of readily machine-readable text in large volumes provides NLP researchers with a steady stream of data that feeds into its currently dominant 'data-bashing' approach. On the other hand, the kind of user-generated data made possible by social media and web activities has posed new challenges: the free-form nature of language on social media often involves spelling inconsistencies, free-form adoption of new terms, and regular violations of grammar norms (Baldwin, 2012). Social media data are highly different from the kind of structured text that NLP historically relied on, such as journals, magazines, newspapers, and books; they tend to be short, informal, unstructured, and dynamic, making it difficult to adapt an NLP tool to the specific social media domain or even rely on context to disambiguate the content. Baldwin (2012) suggested that multimodal content such as images or videos included in social media posts, for example, could as well be used as the rich context against which NLP can draw on disambiguate content; NLP models, therefore, can also draw on various non-textual data sources on social media to enhance their robustness and accuracy. In any case, NLP methods have continued to adapt to working with social media data in order to help researchers extract useful information and answer interesting research questions.

Monitoring and analysing the rich and continuous flows of user-generated content could allow researchers to explore research questions that were not previously possible with static, intermittent forms of data collection such as survey research. Social media data can be collected with certain set parameters (e.g., cut-off dates, identified communities, selected platforms, selected hashtags) and stored in databases or as text files – so that they can be made ready for NLP analysis retrospectively. This is the approach most researchers in the digital humanities currently employ. Social media data can also be collected and processed in real-time for pre-defined purposes; Fine et al. (2020), for example, assessed population-level symptoms of anxiety, depression, and suicide risks in real-time by monitoring Twitter data across healthcare professionals and a community sample group in the US for the first six months of 2020. In so doing, they attempted at estimating the impact of various national events on population mental health by comparing measures of average anxiety, depression, and suicide risk before and after each event through defined periods such as 'early lockdown', 'mid lockdown', and the murder of George Floyd. While it is clear that real-time monitoring of Twitter data during times of crisis (and in this example, during a time when people intensified their online activities due to physical distancing mandates) presents one way to estimate population-level metrics coming out of highly formalised biomedical constructs, it should be noted that this approach is always complementary, rather than replacing, more established methods. There are inherent limits to the representativeness of Twitter data; Blank (2017) found that British Twitter users are younger, wealthier, and better educated than other Internet users, who in turn are younger, wealthier, and better educated than the offline British population. American Twitter users are also younger and wealthier than the rest of the population, but they are not better educated. Twitter users are disproportionately members of elites in both countries. Twitter users also differ from other groups in their online activities and their attitudes. Blank (2017) concluded that the unrepresentative characteristics of Twitter users suggest that Twitter data are not suitable for research where representativeness is important, such as forecasting elections or gaining insight into attitudes, sentiments, or activities of large populations. Apart from issues to do with representativeness, Twitter users are predominantly from the US – accounting for more than 70% of Twitter's overall user base as of November 2022 (Dixon, 2022). Researchers of various diasporas might not find Twitter suitable for their research needs simply because their communities of interest are not there.

The closing off of Facebook's Graph API (application programming interface) since the Cambridge Analytica scandal[1] in 2018 has created significant setbacks for researchers working with Facebook data to trace the dynamic workings and evolution of the communities and groups that organise themselves on this platform. An API is a software intermediary that allows two applications to talk to each other; in the case of Facebook, the Graph API is the primary way to get data into and out of the Facebook platform. The Graph API allows apps to programmatically query data, post new stories, manage ads, upload photos, and perform a wide variety of other tasks. Upon limiting access to the Graph API, Facebook has created a Researcher API to sanction only research that Facebook (now Meta) deems important, API-enabled social research on this platform is now restricted to studies that deal with elections and democracy – an extremely limited response to the call for transparency and accountability by the research community at large. While a large number of diasporic communities are organised on Facebook and as such access to data about these communities would be of significant interest to diaspora researchers, this research program does not currently fall within Meta's approved research agenda. As digital diaspora researchers devise *ad hoc* ways to collect data from these social media platforms in response to the volatile nature of social media platforms, we need to always be cognizant of the larger picture of the digital research landscape and be forthright about what can and cannot be answered with our approaches – as well as being highly flexible and complementary in our method design. The case study provided at the end of this chapter will demonstrate how digital diaspora research can adapt to the changing landscape of digital platforms, as well as the caveats that come with doing digital research by working with automated methods. In the next section, we will look at the typical workflow of NLP applications while considering some of the structural inequities that exist in NLP capabilities with regards to research that involves non-English corpora.

NLP workflow and structural inequities for non-English research

A typical NLP workflow involves two main phases: data pre-processing and data analysis. In data pre-processing, the goal is to turn input texts (e.g., Facebook posts and comments, tweets, and replies) into data structures representing parsed texts that can be used for NLP analysis. The step that gets us from input to output is often called an NLP pipeline; Figure 3.1 represents a typical NLP pipeline for data pre-processing for English-language corpora. While the pipeline in Figure 3.1 starts with sentence segmentation, it is worth noting that not all pipelines necessarily start with this step even when we work with English-language corpora; some steps can sometimes be skipped entirely. Which step to start with first is a decision that varies depending on researcher needs, as well as on particular implementations of different NLP libraries. An example of this is the spaCy library, which implements sentence segmentation much later in the pipeline using the results of the dependency parse.

Sentence segmentation refers to the automated task of breaking down texts into separate sentences. The assumption behind this step is that each sentence should contain at least one stand-alone idea, and that breaking down paragraphs into individual sentences would allow researchers to disaggregate individual ideas from a cluster of ideas, and thus simplify the complexities of their corpus. NLP pipelines usually implement this by splitting apart sentences whenever there is a punctuation mark; more advanced techniques can also be used to ensure that sentences can be identified even when a document is incorrectly formatted. This step is usually not crucial when working with social media data, which tend to be short and poorly punctuated. Each tweet, for example, contains roughly 280 characters as of 2022 and tends not to be written as a structured collection of formal sentences.

Tokenisation refers to the automated breaking down of sentences into individual words – also called tokens. With the English language, this task can be achieved quite straightforwardly by splitting apart words whenever there is a space between them; in this step, punctuation marks can also be treated as separate tokens since punctuation also has meaning. This task can be the most challenging task, however, with monosyllabic languages such as Vietnamese. In Vietnamese, words predominantly consist of a single syllable; while some monosyllabic words have meanings in themselves (e.g., *thích* means to like), a large number of words are made up of monosyllabic types (e.g., *giải thích* means to explain; *thích nghi* means to adapt). About 85% of Vietnamese word types are composed of at least two syllables, and more than 80% of syllable types are words by themselves (Thang et al., 2008; Le et al., 2008). While tokenising by space works well for English, it does not work for languages like Vietnamese. VnCoreNLP – a state-of-the-art NLP

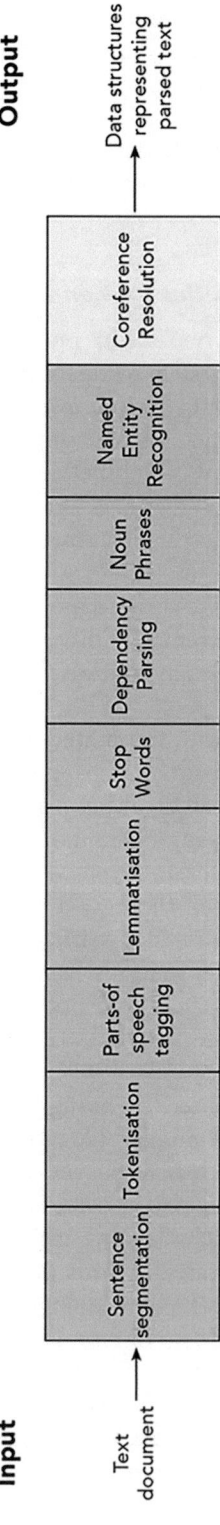

FIGURE 3.1   A typical NLP pipeline for the English language.

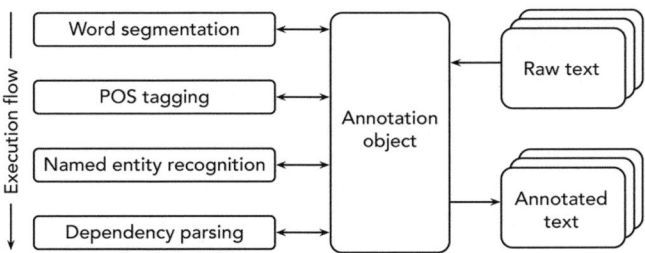

FIGURE 3.2 The VnCoreNLP pipeline for the Vietnamese language.

pipeline for the Vietnamese language – used a transformation rule-based learning model for Vietnamese word segmentation by automatically constructing a Single Classification Ripple Down Rules (SCRDR) tree (Compton & Jansen, 1990) to correct wrong segmentations given by a longest matching based word segmenter (Nguyen et al., 2018). Lesser-resourced languages such as Vietnamese tend to rely on feature-based models in their NLP pipelines, rather than the state-of-the-art network-based models, whose speed and accuracy are contingent on massive amounts of training corpora. VnCoreNLP itself also has a distinct workflow compared to that of a typical English-language workflow; Figure 3.2 illustrates VnCoreNLP workflow, with word segmentation designed to be the first step while skipping sentence segmentation altogether.

Part-of-speech (POS) tagging refers to the automated process of predicting a token's part of speech – whether the token in question is a noun, verb, adjective, adverb, interjection, conjunction, and so on. Different languages will naturally have different POS tagsets; Table 3.1 lists the Penn TreeBank tagset for the English language and Table 3.2 lists the VnCoreNLP tagset for the Vietnamese language. POS taggers deploy complex models that rely on annotated training data to learn the probabilities of a specific POS along with a range of other parameters; deep learning POS taggers can also be built to directly transform sequences of words into sequences of POS taggers.

Stop word identification refers to the removal of noise in a dataset with the help of automated methods in an NLP pipeline. In the English language, filler words that appear very frequently such as 'and, 'the', and 'a' become noise when we conduct statistical analysis on texts, as they appear a lot more often than other words. Some NLP pipelines will flag these as stop words so that the researcher might want to filter them out before doing any statistical analysis. Stop words are usually identified by checking a hardcoded list of known stop words in a particular language. For social media data, a relevant and related task is **text normalisation** – where researchers use a dictionary of known correctly spelt terms to identify orthographic errors in an input text and then correct them. Basic normalisation deals with errors detected at POS tagging such as unknown words and misspelled words, as well as words that are products of internet culture (e.g., LMFAO, LOL, ROFL). Advanced normalisation can be more flexible and take a supervised automatic approach trained on an external dataset (Farzindar & Inkpen, 2020).

Dependency parsing refers to the automated process of determining how tokenised words relate to each other in a sentence. A dependency parser extracts pairs of words that are in a syntactic dependency relation; their relations can be verb-subject, verb-object, noun-modifier, and so on (Farzindar & Inkpen, 2020). Dependency parsers can be built

TABLE 3.1 Penn TreeBank tagset

Number	Tag	Description
1	CC	Coordinating conjunction
2	CD	Cardinal number
3	DT	Determiner
4	EX	Existential there
5	FW	Foreign word
6	IN	Preposition or subordinating conjunction
7	JJ	Adjective
8	JJR	Adjective, comparative
9	JJS	Adjective, superlative
10	LS	List item marker
11	MD	Model
12	NN	Noun, singular or mass
13	NNS	Noun, plural
14	NNP	Proper noun, singular
15	NNPS	Proper noun, plural
16	PDT	Predeterminer
17	POS	Possessive ending
18	PRP	Personal pronoun
19	PRP$	Possessive pronoun
20	RB	Adverb
21	RBR	Adverb, comparative
22	RBS	Adverb, superlative
23	RP	Particle
24	SYM	Symbol
25	TO	To
26	UH	Interjection
27	VB	Verb, base form
28	VBD	Verb, past tense
29	VBG	Verb, gerund or present participle
30	VBN	Verb, past participle
31	VBP	Verb, non-3rd person singular present
32	VBZ	Verb, 3rd person singular present
33	WDT	Wh-determiner
34	WP	Wh-pronoun
35	WP$	Possessive wh-pronoun
36	WRB	Wh-adverb

from rule-based approaches, probabilistic models, or deep learning using neural networks. An example of dependency parsers is Parsey McParseface – an open-sourced machine learning parser developed by Google based on Tensorflow. Parsey McParseface recovers individual dependencies between words with over 94% accuracy; Google claims that it is the world's most accurate syntactic parser (Petrov, 2016). VnCoreNLP produced around 70% accuracy with their dependency parser when tested on the last 1020 sentences of a benchmark Vietnamese dependency treebank called VnDT (Nguyen et al., 2014).

Named entity recognition (NER) refers to the detection and labelling of nouns with names – usually of person, organisation, and location – by detecting the boundaries of these phrases. NER can be done based on linguistic grammar or using statistical methods.

TABLE 3.2 The VnCoreNLP tagset

Number	Tag	Description
1	Np	Proper noun
2	Nc	Classifier noun
3	Nu	Unit noun
4	N	Noun
5	Ny	Abbreviated noun
6	Nb	(Foreign) borrowed noun
7	V	Verb
8	Vb	(Foreign) borrowed verb
9	A	Adjective
10	P	Pronoun
11	R	Adverb
12	L	Determiner
13	M	Numeral/Quantity
14	E	Preposition
15	C	Subordinating conjunction
16	Cc	Coordinating conjunction
17	I	Interjection/Exclamation
18	T	Particle/Auxiliary, modal words
19	Y	Abbreviation
20	Z	Bound morpheme
21	X	Un-definition/Other
22	CH	Punctuation and symbols

Supervised learning techniques are used in cases where there is an availability of annotated training datasets, where named entities are tagged by human annotators. In the case of VnCoreNLP, contiguous syllables constituting a full name are merged to form a word and POS taggers are trained to assign a label to the entire full name (e.g., Nguyễn_Hồng_Hải_Đăng would be correctly tagged as Np). **Coreference resolution** can also be used to detect the noun that a pronoun refers to in order to detect different noun phrases that refer to the same entity (Farzindar & Inkpen, 2020). As such, coreference resolution as a task supports NER as much as it does advances the parsing of texts to extract useful information. Coreference resolution is considered one of the most difficult steps to implement in an NLP pipeline for languages such as English; as we see from Figure 3.2, however, this step is not relevant to NLP pipelines for the Vietnamese language, for example.

After running input data through an appropriate NLP pipeline, the output should be correctly segmented and tagged tokens in a format that can be used for NLP analysis. In the context of social media research, the most frequently conducted NLP analyses are topic modelling and sentiment analysis. **Topic modelling** refers to a method for unsupervised classification of texts, which slots texts into groups (understood as topics) when we are not sure what we are looking for. This is particularly useful in contexts where the empirical domain of interest is esoteric and/or understudied. The case study included at the end of this chapter performs topic modelling following the Latent Dirichlet Allocation (LDA) method on data collected from traditional medicine groups on Facebook – which is an understudied empirical domain within digital research. LDA is a popular probabilistic method (Blei et al., 2003; Blei & Mcauliffe, 2007) to explore a large corpus of text when researchers are not sure what they are looking for, because the assumptions in LDA are robust and well-understood. LDA defines topic as a distribution of words over a fixed vocabulary – the total amount of segmented words in a dataset, for example – and assumes that a 'topic' can be understood as a collection of words that have different probabilities of appearance in particular contexts discussing the topic – which could be a discussion thread in the context of social media data. While topics cannot be directly observed, topic modelling is a way of extrapolating backwards from a collection of documents to infer the topics that could have generated them. LDA assumes that the hidden probabilistic structure of word distribution within the dataset resembles the thematic structure of the dataset; LDA as a method does not capture the correlation among topics. Cousins of LDA include Topics Over Time (TOT) modelling – which captures not only the low-dimensional structure of data, but also how the structure changes over time (Wang & McCallum, 2006); dynamic topic modelling – which deploys a family of probabilistic time series models to analyse the time evolution of topics in large document collections (Blei & Lafferty, 2006); hierarchical LDA – which employs a stochastic process that assigns probability distributions to ensembles of infinitely deep, infinitely branching trees (Blei et al., 2010); and Pachinko allocation – which captures arbitrary, nested, and possibly sparse correlations between topics using a directed acyclic graph (Li & McCallum, 2006). Blei and Lafferty (2007) also developed a correlated topic model (CTM) to address LDA's inability to detect correlations among topics and allow researchers to examine relationships between topics; this method is most useful when researchers identified a large number of topics or become informed either by theory or by their engagement with the corpus that a lot of the topics within their corpus are going to be related.

Sentiment analysis, also known as opinion mining, refers to the automated process of identifying and extracting the overall polarity of a text – whether it is generally positive, negative, or neutral (Pang & Lee, 2008). The polarity approach is useful to commercial domains to gauge the overall sentiments associated with certain products through aggregated analysis of product reviews, restaurant reviews, or movie reviews, for example. Sentiment analysis can also engage with more advanced techniques to discern emotional states such as enjoyment, anger, disgust, sadness, fear, and surprise; the assumption behind this approach is that patterns in text can be quantified towards informing researchers about a person's psychological state based on analysis of their verbal behaviour. The methods used in semantic analysis can be corpus-based using machine learning (where a set of training data is labelled by humans – also known as supervised learning), rule-based using sentiment lexicons/dictionaries (using unlabelled data – also known as unsupervised learning), or a mix of both approaches. In rule-based methods, lists of emotion words are compiled so that term-counting features can be deployed to place emotions on a scale. Some examples of such emotion lexicons are ANEW (Bradley & Lang, 1999), WordNetAffect (Strapparava & Valitutti, 2004), Linguistic Inquiry and Word Count – LIWC (Tausczik & Pennebaker, 2010), and SentiWordNet (Baccianella et al., 2010) for the English language. With machine learning techniques, probabilistic algorithms are devised to assign a probability that a given word or phrase should be considered positive or negative (Naïve Bayesian classifier); linear regression algorithms are devised to show relationships between the text input and the polarity output to then determine where words and phrases fall on a scale of polarity from positive to negative. Similar to the linear regression approach but a lot more advanced is the Support Vector Machines (SVM) approach, where algorithms are devised to train and classify texts within a sentiment polarity model and where each data point is plotted in n-dimensional space (in which n is the number of features) with the value of each feature being the value of a particular coordinate. Classification with SVM is then performed by finding the hyper-plane that differentiates the classes well; as such, SVM represents a multi-dimensional approach to sentiment analysis where diverse sources of potentially pertinent information are brought together to improve accuracy (Mullen & Collier, 2004).

Sentiment analysis is used in the social media context for a range of applications. Working with Twitter data, for example, Wang et al. (2012) monitored shifts in sentiment regarding political candidates in the 2012 US presidential election; Thelwall et al. (2011) monitored 30 significant US-based cultural events on Twitter and found that popular events are normally associated with increases in negative sentiment strength while identifying some evidence that peaks of interest in events have stronger positive sentiment than the time before the peak. Working with Facebook data (prior to the events of Cambridge Analytica), Ortigosa et al. (2014) devised a sentiment analysis method specifically tailored for Facebook and developed a Facebook app called SentBulk, which retrieves messages written by users on Facebook and classifies them according to their polarity before showing the results to the users through an interactive interface. The ultimate application of this project was to aid students and teachers in online learning to adapt e-learning systems to support personalized learning by using the identified emotional state of users as course recommendation feedback. For researchers of diasporas, understanding community-level sentiments around specific time periods or cultural events could open up interesting avenues of research, especially when the automated tracing of these dynamic sentiments is accompanied by fine-grained ethnographic studies into what happens within the community from the bottom up.

There are currently structural inequities in sentiment analysis capabilities across different languages, with the majority of the resources available catered to the English language. Sentiment analysis in multiple languages is often addressed by transferring knowledge from resource-rich to resource-poor languages (Dashtipour et al., 2016); one approach is to translate non-English texts into English with machine translation and then deploy existing English-language resources, but this approach carries various issues to do with sparseness and data noise, as well as issues to do with essential parts of a text not being translated due to the nature of good translation not done in a literal word-for-word fashion – which results in loss (Denecke, 2008). Multilingual lexical resources specific to sentiment analysis are being developed, such as the NTCIR corpus of news articles in English, Chinese, and Japanese (Seki et al., 2010), the Chinese sentiment corpus ChnSentiCorp (Tan & Zhang, 2008), or the Czechian sentiment dataset developed by Habernal et al. (2014) which includes Facebook comments, movie reviews, and product reviews. Nevertheless, pre-training NLP models from scratch using large corpora required for meaningful NLP results is expensive; when a team can download a pre-trained model, this overhead can be avoided, but this is only possible if such a pre-trained model is available in their language (Wali et al., 2020).

Current disparities within NLP resources lie in the state that NLP pre-trained models only exist for a few languages, and that many authors and older (thus more prestigious) language conferences have been focusing on already resource-rich languages with little focus on lesser-resourced languages (Wali et al., 2020). These disparities are not unique to NLP, however; at the structural level, close to 98% of the world's living languages are digitally disadvantaged – meaning they are not supported on the most popular devices, operating systems, browsers, and mobile applications (Unicode, 2015). Kornai (2013) estimates that at most 5% of the 7000+ languages in use will achieve 'digital vitality', while the other 95% would face 'digital extinction'. Digitally disadvantaged languages face a range of challenges in the digital context, including gaps in equitable access, negative impacts on the integrity of these languages, scripts, writing systems, and knowledge systems, and vulnerability to harm through digital surveillance and under-moderation of language content (Zaugg et al., 2022). Researchers of diaspora often find themselves working with diverse linguistic communities whose languages are overlooked in the current landscape of NLP research; there is also currently little institutional incentive for diaspora researchers to engage with computationally intensive research even within digital humanities programs due to gaps in the accessibility and availability of NLP resources. As a result of this, there is much untapped potential for diaspora researchers to engage with computational methods by working with colleagues from across disciplines, developing interdisciplinary and cross-disciplinary research programs, and supplementing existing shortcomings in NLP capabilities with more established and complementary methods in the qualitative tradition. In the case study below, I will demonstrate how even with currently limited NLP capabilities for a lesser-resourced language such as Vietnamese, diaspora researchers can still ask and answer interesting and innovative questions by working with Facebook data.

Case study: topic modelling and online discourse about traditional medicine

This section reports on the methods and summarises the key findings in Nguyen (2021b) as a case study for marrying NLP and social network analysis to studying online diasporic discourse. In particular, this section introduces the domain-specific literature that the paper contributes to by briefly introducing its theoretical framework and its research questions before reporting in detail on the practical methodological choices made in its analysis. We consider some aspects of social network analysis that were not included in Chapter 2 in this case study; the case study also demonstrates how NLP can help researchers explore a large social media dataset in ways that do not require a manual coding scheme.

Literature overview & research questions

Traditional medicine is hugely popular throughout Southeast Asia and other parts of the world. The development of the internet and online social networks in these contexts has enabled a significant proliferation of non-biomedical knowledge and practices via platforms such as Facebook. People use Facebook to advocate for non-biomedical alternatives to unaffordable biomedicine, share family medical recipes, discuss medicinal properties of indigenous plants, buy and sell these plants, and even crowdsource disease diagnoses. This case study examines the network characteristics of, and discourses present within, three popular Vietnamese non-biomedical knowledge Facebook sites over a period of five years. These large-scale datasets are studied using social network analysis and generative statistical models for topic analysis (Latent Dirichlet allocation).

Throughout history, knowledges produced by these different social classes received different levels of marginalisation, under French colonialism, and through competition from the more 'learned' and established Chinese medical traditions. Thuốc Bắc (Northern medicine), for example, is commonly associated with the literati class and is heavily influenced by Chinese medicine, whereas thuốc Nam (Southern medicine) is commonly associated with medical families (Monnais et al., 2011). Those who make a living by scouting, growing, collecting, prescribing, and selling raw medicinal plants also contribute to this knowledge ecosystem with their own interpretations and revisions of family recipes through direct interaction with patients as well as experience with local flora and fauna. The propagation of these knowledges throughout history has followed flexible patterns and structures that enabled composition, retention, reperformance, as well as constant revision. These knowledges fulfill clear and immediate functions for communities that maintain them – namely managing illnesses and preserving health – through their ability to vary and respond to different circumstances.

This case study is interested in exploring whether the democratisation of these knowledges through decentralised propagation on social media has changed the very fabric of their sociality: whether new crossovers and contacts are being forged as a result of intensified and increasingly visible flows of these historically marginalised knowledges. What also remains unknown is the content of knowledges being exchanged on these sites, as well as its associated discourses. These unknowns may be expressed in this way:

RQ1 What are the network characteristics of Vietnamese non-biomedical sites on Facebook?
RQ2 What types of non-biomedical knowledge discourses are present within these sites?

Methodological choices and motivation

This case study uses social network analysis (SNA) and machine learning techniques (particularly NLP and topic modelling) to answer the two research questions. Social network analysis is both a set of theoretical perspectives and analytic techniques used to examine how exchanges between individual units both shape, and are shaped by, the larger context in which those two individual units are embedded (Carolan, 2016). SNA assumes an emphasis on relations among individuals and not their individual attributes, with particular focus not on individuals as members of discreet groups but rather as members of overlapping networks (Marin & Wellman, 2011; Carolan, 2016). Formal network measures provide a rigorous language with which to discern network properties and make sense of the way non-biomedical knowledge propagates in internet environments (Hanneman & Riddle, 2016).

Machine learning techniques, particularly that of an unsupervised nature such as Latent-Dirichlet Allocation (LDA), allow for the statistically driven uncovering of topics 'hidden' in the dataset under the assumption that underlying topics match with the probabilistic distribution of words over a set vocabulary (Blei et al., 2003; Blei & Mcauliffe, 2007). These techniques allow for robust and automated discovery of a large corpus, which is useful for the current context. Subsequent interpretation and labelling of the topics discovered by this automated process are conducted by the researcher, which ensures that these topics are meaningful according to human evaluation standards.

Site selection strategy

The selected sites were purposively sampled from an automated list of 1900 Vietnamese non-biomedical health groups and pages on Facebook, based on three criteria: (i) popularity, measured in number of active participants, (ii) activity, measured in number of posts per week, and (iii) privacy settings, in that only public sites with fully public content are selected. Criteria i and ii ensure the sites sampled are active rather than abandoned sites. Criterion iii ensures that the automatic collection of textual data does not violate participants' privacy; informed consent was not sought because participants were engaged in a public discussion and no personally identifiable information was collected. These criteria are also consistent with what van Dijck and Poell (2013) theorise as the four grounding principles of social media logic: programmability (the mutual layering of technological features and human agency in shaping platform usage), popularity (the algorithmic and socioeconomic conditioning of influence and importance), connectivity (socio-technical affordances of the platform apparatus that mediates user activity), and datafication (the ability of network platforms to render into data aspects of life that were not quantified before). Only sites with over 30,000 members with a posting frequency of over 10 posts per day were selected for the sample. The automated list was generated by automating searches using the search function on Facebook with 21 different keywords. Table 3.1 provides descriptive statistics of the sampled sites.

The sites sampled here are public sites where membership is not moderated, as opposed to moderated membership where applicants are required to answer a set of questions pending approval of site administrators. As such, gatekeeping within these sites is minimal. Sites 1 and 2 are more similar to each other than they are to Site 3, in that they are both sites built exclusively around promoting and sharing Vietnamese traditional medical recipes. Site 1 centres on the sharing of traditional Vietnamese medicine in general, while Site 2 is focused on Southern medicine and family recipes. Site 3 has an explicit anti-biomedicine philosophy; the site description outlines its advocacy against overreliance on biomedicine as an expensive therapeutic option. Each of these sites corresponds to a different existing knowledge paradigm that characterises the diversity of non-biomedical practices in Vietnam. Together these sites are the top active sites for the exchange of non-biomedical knowledge on Facebook in Vietnam.

Data collection and network generation

Data were collected with a purpose-built web scraper. The scraper utilises the Puppeteer library developed by Google to collect publicly available data via the Chrome web browser. The scraper collected all text content on original posts and their associated comments over five years, from 19 August 2014 to 19 August 2019. As shown in Table 3.3, the dataset contains 7,957 unique posts and 41,032 comments,

TABLE 3.3 Descriptive statistics of sampled sites (as of August 2019)

Site	Number of members	Number of posts	Number of comments	Number of 'reacts'	Number of 'shares'
Site 1 Good traditional medical recipes (*Các bài thuốc dân gian hay*)	38,744	3,940	16,459	55,511	13,921
Site 2 Southern medicinal plants and family recipes (*Cây thuốc nam và những bài thuốc gia truyền*)	82,008	2,983	6,874	36,159	69,120
Site 3 Your wise medical cabinet (*Tủ thuốc thông thái*)	45,829	1,034	17,699	94	29
Total	166,581	7,957	41,032	91,764	83,070

representing the activities of 166,581 unique members. Table 3.3 also includes the total number of 'shares' across all posts for each of the three public sites. A 'share' means that a viewer of a post has shared the post and any associated links within their personal network. All together for this dataset, posts were shared over 80,000 times. A 'react' on Facebook is an emotional response that takes on one out of six available emotional reactions, signified by six distinct emojis. Due to the limits of non-API scraping, finer data for the 'react' construct are not available. There were altogether 91,764 reacts in this dataset. Although site 3 has a significant number of members and a high count of commenting activity, their 'React' and 'Share' metrics are significantly lower than sites 1 and 2. This peculiar dynamic could be due to the nature of site 3 as a consumer movement site; while the positive framing of knowledge propagation found in sites 1 and 2 might entice more affective engagement and sharing behaviour, the negative framing of site 3 (anti-biomedicine) might limit these forms of engagement. Table 3.4 provides descriptive statistics for the co-commenting networks constructed from the data collected, with density measuring the prevalence of dyadic linkage or direct tie within a social network (Frey, 2018). The index of network density is expressed as the ratio of observed ties (edges) to all possible pairwise ties in a network, whose value ranges between 0 and 1. It can be interpreted as the proportion of potential ties that are actually present (Frey, 2018).

TABLE 3.4 Descriptive statistics for constructed co-commenting networks

Site	Nodes	Edges	Density
1	3560	10208	0.000805
2	2783	5136	0.000663
3	5475	12137	0.000405

Results – network analysis

From the data collected, undirected co-commenting activity networks were constructed for each site, wherein nodes represent unique users and edges represent comments. An undirected edge is created if users co-commented on any posts; a low bar is therefore set for creating a relationship between users. Two users are connected if they have both commented on the same post within the five-year time period. This construction adapted the methods proposed by Graham and Ackland (2016), as data on 'Reacts' and 'Shares' could not be collected with the current non-API scraper.

User comment networks provide a more fine-grained understanding of the structure and nature of user interactivity on Vietnamese non-biomedical knowledge sites, even though Facebook does not directly provide this type of data. Comments are individually composed as open-text and as such represent a novel contribution and extension to discussion, as compared to a reproduction of a previous contribution through reacting or sharing. Comments also contribute differently to the propagation of non-biomedical knowledge on Facebook, as users often read each other's comments, interpret and learn from them, as well as engage in discourse by posting their own comments. Within online health contexts, users actively seek and provide various types of social support by interacting with each other via the interface of social media platforms (Hether et al., 2016; Nguyen, 2021a). As such, these networks provide interesting insights into user (co)participation and discourse dynamics not only within each site, but also across these sites as an aggregate network. Figure 3.3 presents a sample discussion thread from the dataset.

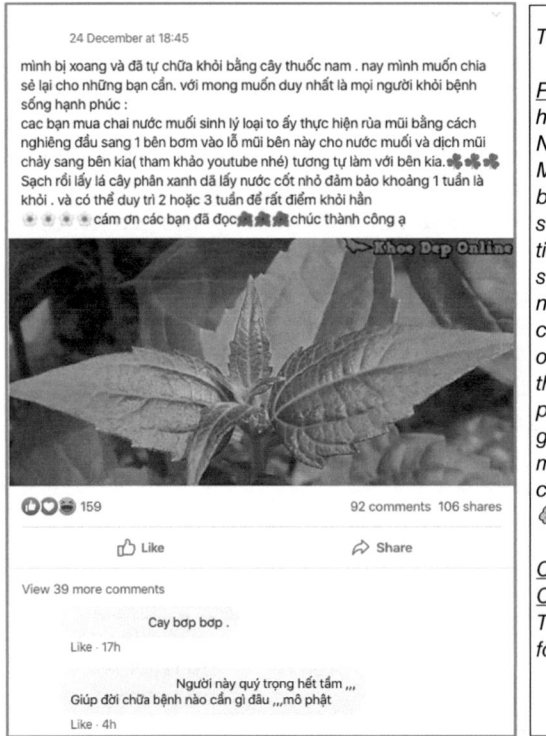

Text translation:

Post content: " I used to have sinusitis and I had it cured by southern medicine plants. Now I want to share to anyone that needs it. My only hope is that everyone will no longer be sick and be happy: Buy a big bottle of saline solution then perform nose washing by tilting your head to one side. Pumping the solution into one nostril until the solution and nasal fluid flow to the other one (please also consult Youtube) then do the same with the other nostril 😊😊😊. Once it's clean, pound the green manure leaves to get its extract and pump it into your nose. I guarantee it will be gone after around one week. And you can maintain for two to three weeks for it to be completely gone.😊😊😊 Thanks for reading 😊😊😊 Good luck. ";

Comment 1:" But this is Siam weed ";
Comment 2: "This person is a treasure… They are giving medical help without asking for anything return… Namo amitabud.")

FIGURE 3.3 Sample Facebook thread from the dataset (usernames are hidden).

TABLE 3.5 Out-degree distribution of user activity by site-level network

Network	Number of users who commented twice or less	Percentage of total activity within network
Site 1	2082	55.24%
Site 2	1606	54.66%
Site 3	3904	70.64%

Although a large number of users are members of these sites, only a small portion of users participated in the form of commenting regularly. As Table 3.5 shows, a majority of users only commented on a post once or twice over a five-year period, which constitutes a highly skewed out-degree distribution. Only a small subset of users within each site contributes in terms of posting content and commenting frequently, and there is a 'long tail' of users who are very infrequent in their commenting activity.

On an aggregate network level, it is of interest to examine whether users who participate in one site-level network also participate in other sites within the sample. Perhaps quite surprisingly, the percentage of cross-participation within the selected sites is very low (see Table 3.6). In each pair of the site-level network, cross-participation is below one per cent, with a particularly low cross-participation rate between sites 2 and 3. As such, not only does site 3 have a very low effective engagement rate ('Reacts') and sharing behaviour within the site itself, its members also do not seem to engage in the other most popular non-biomedical knowledge sites. Contrasting this against the high intensity of discussion that happens within the site (high counts of comments and posts), it seems that non-biomedical knowledge groups that rely on an anti-biomedicine philosophy could be exhibiting cult-like behaviour, in the sense that content is frequently discussed within, but does not propagate outside of, the group. This implies a sense of insularity that does not benefit knowledge exchange among the examined groups. This supports the observation of the 'transient user' on Facebook, wherein users 'pass by' discursive groups without investing in maintaining discursive relationships or co-ordinating different discourses across groups.

Considering that each network from the sampled sites remains largely separate from each other, it is worthwhile to further examine whether these networks exhibit the properties of 'small world' networks (Watts & Strogatz, 1998). Small world networks are network structures that are both highly locally clustered and have a short path length – two network characteristics that are usually divergent (Watts, 1999). Small world networks are interesting for many reasons. For example, small world networks enable infectious diseases to spread much more quickly and easily than other types of networks, as the dynamics of the network is an 'explicit function of structure' (Watts & Strogatz, 1998, p. 441). Empirical research has also shown

TABLE 3.6 Percentage of user participation across pairs of site-level networks

	Site 1	Site 2	Site 3
Site 1	100%	0.83%	0.46%
Site 2	0.83%	100%	0.06%
Site 3	0.46%	0.06%	100%

that the more a network exhibits characteristics of a small world, the more connected actors are to each other and connected by persons who know each other well through past interactions, or through having had past interactions with common third parties (Uzzi & Spiro, 2005). These conditions allow information circulated in separate clusters to also circulate to other clusters, and to gain the credibility that unfamiliar materials require to be regarded as valuable in new contexts and subsequently used by other members of other clusters (Uzzi & Spiro, 2005). Small world networks are also interesting because they are robust and resistant to damage, in the sense that randomly removing nodes from the network will not significantly impact the effectiveness and dynamics of the network. The small-world phenomenon is not only common in sparse networks with many vertices, as even a tiny fraction of shortcuts would suffice; research has demonstrated that it is common in biological, social, and artificial systems (Watts & Strogatz, 1998; Uzzi et al., 2007; Telesford et al., 2011; Bassett et al., 2017; Opsahl et al., 2017).

Two methods were used to assess whether the three co-commenting networks are 'small worlds'. The first approach follows the conditions set in Watts and Strogatz (1998), where a network is considered small world if (1) its average local clustering coefficient is much greater than a random network generated from the same set of vertices and (2) the mean shortest path length of the network is approximately the same as the associated random network. To do this, I calculated the average local clustering coefficient and mean shortest path length for the three networks studied here and compared these metrics against those of three randomly generated networks with the same number of edge sets. I generated these three random networks using the Erdős-Rényi model implementation in the 'igraph' R package (Csardi & Nepusz, 2006). The second approach employs Humphries and Gurney (2008)'s small-worldness index, where the index is calculated as transitivity (normalised by the random transitivity) over the average shortest path length (normalised by the random average shortest path length). Transitivity, an alternative definition of network clustering, is understood as the propensity for two neighbours of a network node to also be neighbours of one another (Newman et al., 2000; Newman, 2009). Using the 'qgraph' R package (Epskamp et al., 2019), the average of the same indices was calculated on 1000 random networks for each co-commenting network. A network can be said to be 'small world' if its small world index is higher than one; a stricter rule requires the index to be higher than three (Humphries & Gurney, 2008). Results are presented in Table 3.7, where all three networks satisfy the conditions in both approaches to be small worlds.

TABLE 3.7 'Small world' metrics for user co-comment networks vs. random graphs (bolded in brackets) and small-worldness index (Humphries & Gurney, 2008)

Network	Average local clustering coefficient	Average shortest path length	Small-worldness index
Site 1	0.0214 (**0.0018**)	4.414 (**4.871**)	3.353
Site 2	0.0058 (**0.0014**)	5.525 (**6.152**)	9.689
Site 3	0.0099 (**0.0007**)	4.316 (**5.933**)	9.704

Results - text analysis and topic modelling

In order to understand the nature of discourse on these networks, topic modelling was performed on the complete set of textual data collected, including original posts and their associated comments, across all three sites. Probabilistic topic modelling allows for efficient and reproducible analysis of large amounts of textual data without requiring prior annotations or labelling of the textual corpus; topics that emerge from this analysis are determined through the co-occurrence of words and the themes they carry within the texts (Blei, 2012). The analysis was carried out using the LDA method (Blei et al., 2003; Blei & Mcauliffe, 2007), an established generative statistical topic modelling method within the social sciences (DiMaggio, 2015). LDA defines a topic as a distribution over a fixed vocabulary; it assumes that topics are specified before textual data are generated (Blei, 2012). This method formalises the intuition that there exists hidden topics within set texts, and that these hidden topics can be inferred by examining words that appear with particular probabilities. The utility of topic models lies in the property that the hidden structures inferred resemble the thematic structure of the dataset (Blei, 2012).

To prepare the corpus for LDA, a natural language processing annotation pipeline specific to the Vietnamese language was used to segment individual words and tag them with the appropriate part of speech (Vu et al., 2018). The analysis was then conducted on 469,388 noun terms such as 'cancer', 'monk fruit', and 'hibiscus', that occur in at least 80% of 25,356 discussion threads in the dataset. The rationale behind this method is based on the observation that, within this dataset, discussions usually involve support seeking and provision (i.e., people naming a disease or condition to seek out names of medicinal plants or ingredients that supposedly help with said disease or condition). As such, disease names and names of medicinal plants or ingredients that appear alongside each other in the same discussion threads with high frequency could indicate popular non-biomedical therapeutic beliefs and practices. Specifying the LDA model consists of three steps: (1) draw k topics from a symmetric Dirichlet distribution, (2) for each document d, draw topic proportions from a symmetric Dirichlet distribution, and (3) for each word n in each document d, draw a topic assignment from the topic proportions and draw the word from a multinomial probability distribution conditioned on the topic (Grün & Hornik, 2011). There are many approaches to choosing k number of topics, such as perplexity (Blei et al., 2003), marginal likelihood (Griffiths & Steyvers, 2004), density (Cao et al., 2009), and symmetric Kullback–Leibler divergence (Arun et al., 2010). No one approach is currently considered the standard; researchers working with LDA often choose the method most appropriate with the nature of their data. To ensure rigorous k selection, I calculated all four metrics using the 'ldatuning' and 'topicmodels' R packages (Nikita, 2016; Grün & Hornik, 2011). Figures 3.4 and 3.5 plot the results of these metrics. Figure 3.4 indicates that the best number of topics lies somewhere in the range between 70–160, while Figure 3.5 indicates that the range is between 60–80. It is documented that Cao et al. (2009) and Arun et al. (2010) metrics tend to overfit the data (Hou-Liu, 2018; Gerlach et al., 2018). Marginal likelihood (Griffiths & Steyvers, 2004) has been widely used as a measure to specify k on large-scale social media datasets across different languages and health topics, where the topic candidate with the highest likelihood value is considered the best fit. Perplexity is often used alongside marginal likelihood as a method of cross-validating k selection, where lower perplexity is considered a better fit (Hoang, 2015). Based on these analyses, k topic is selected at 70.

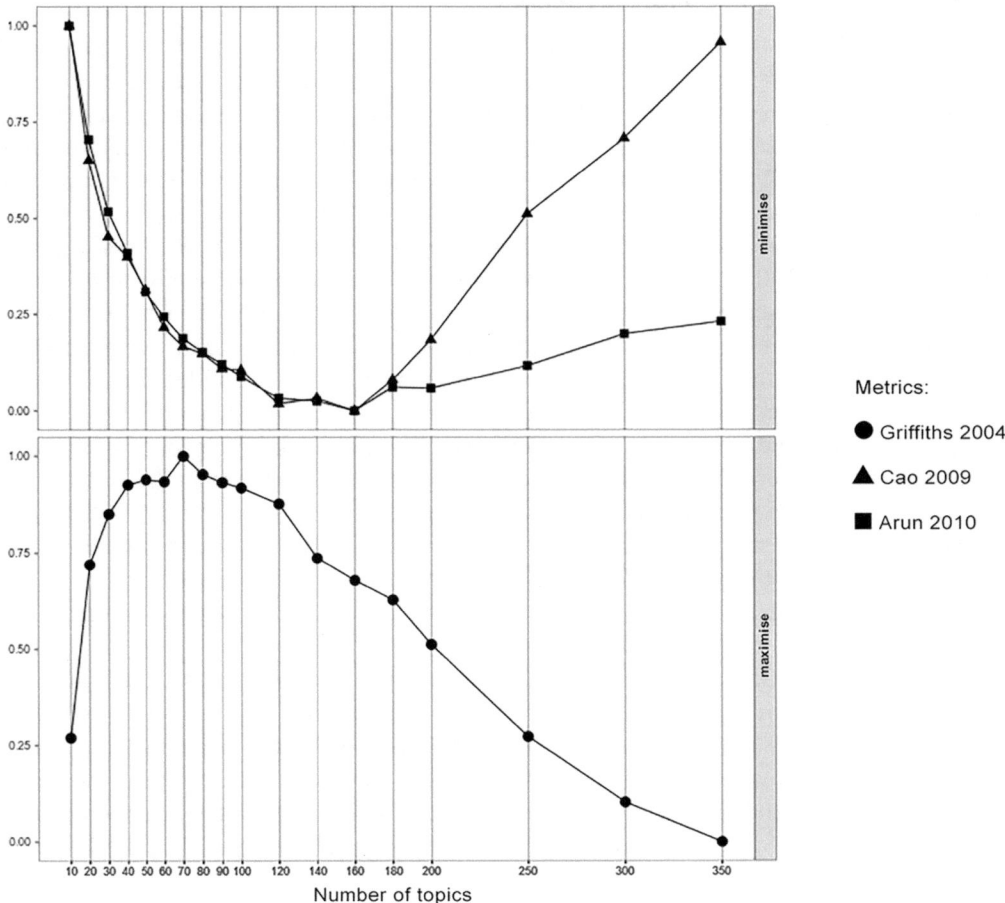

FIGURE 3.4 Computed metrics and estimated k number of topics using the 'ldatuning' R package.

To validate the topic model fitted to the current data, Maier et al. (2018) suggested employing 'systematically structured combinations of existing metrics and in-depth investigation to boost the significance of the validation process' (p. 97). They devised a three-step process to operationalise this: summarising the most important quantitative information from the model, outlining exclusion strategies for uninterpretable topics, and close reading of the data and labelling of a topic.

Maier et al. (2018) proposed the use of four particular metrics: rank-1 metric (Evans, 2014), coherence (Mimno et al., 2011), relevance (Sievert & Shirley, 2014), and the Hirschman-Herfindahl Index (HHI). Rank-1 metric is useful for helping identify background topics. Coherence score, when applied to single topics, can help guide intuition in interpretation. Relevance score can help reorder the top words of a topic by considering their overall corpus frequency through manipulating the weighting parameter λ, with the best interpretability of topics using a λ-value close to 0.6 (Sievert & Shirley, 2014). Finally, HHI = 1 signifies maximum concentration (the topic is pronounced by only one source) and a very low HHI value, conversely, indicates that a topic can be found in

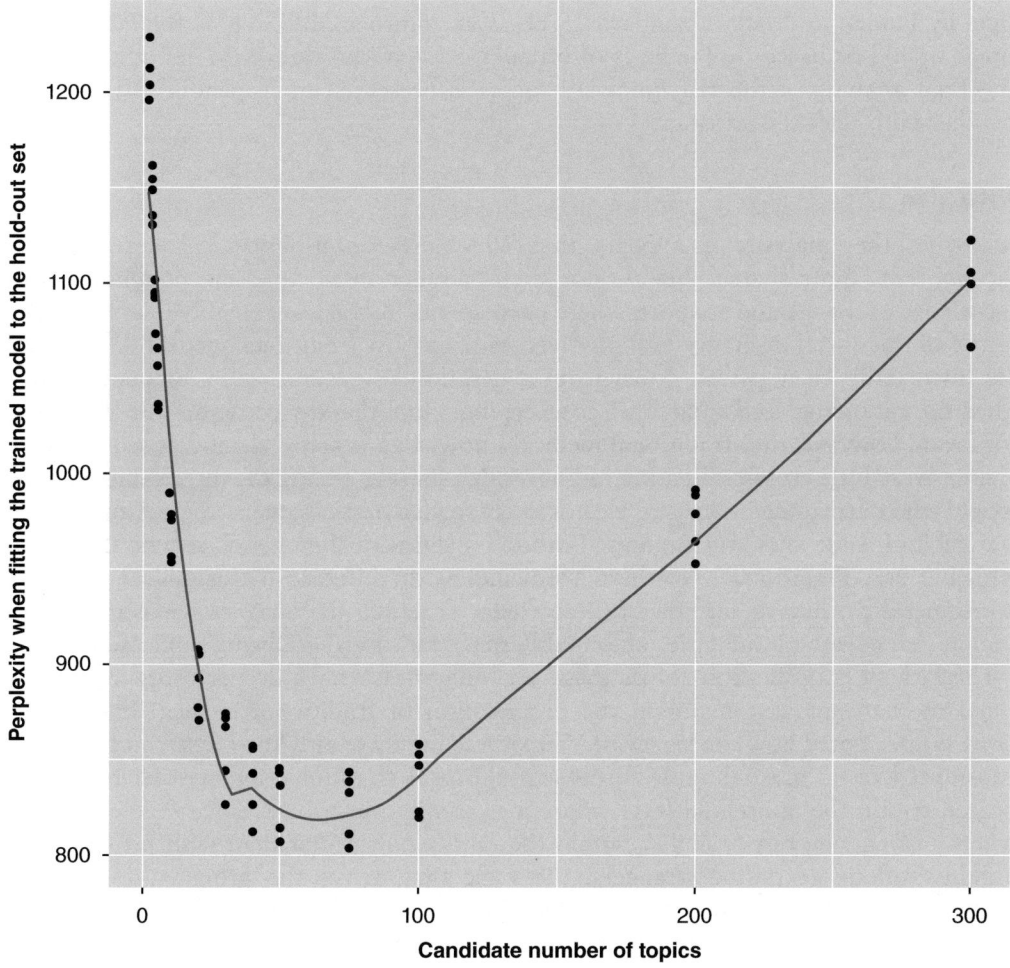

FIGURE 3.5 Computed perplexity and estimated k number of topics using the 'topicmodel' R package.

many sources. HHI, while useful in Maier et al. (2018)'s specific dataset which tracks the hyperlink network of over 300,000 websites, is not useful to the current Facebook dataset. HHI is therefore not calculated here. A sample summary of statistics of these metrics is included in Appendix 2.

Of the 70 topics generated, there were 17 overlapping topics (i.e., topic 2 appeared 17 times in the results). Only four topics include mostly 'junk' terms such as 'shhh', 'kkkkk', 'hehehehe' – which are words that were not interpretable in relation to others in the topic. These terms are generally considered to be an artefact of social media data and the phatic nature of online communication, which is commonly encountered. These four topics, which contain mostly 'junk terms', are also excluded from the analysis. From this filtering process, there are 49 topics that are eligible for analysis. The most representative threads containing each identified topic were retrieved; a close reading of each thread was then conducted manually to ensure that the topic labelling reflects the underlying topic by human evaluation standards. This is an indispensable step as the labelling of topics should be based on the basis of broader context knowledge (Maier et al., 2018). The final analysis of the remaining 49 topics is presented in Appendix 1 Table 3.6, together with relevant statistics.

Discussion

Following the analysis, it appears that Vietnamese non-biomedical networks on Facebook are quite sparse: they do not seem to function as close-knit communities of knowledge exchange and support, where participants interact sustainably over time. The nature of this social exchange pattern diverges from how traditional medical knowledge has historically been passed down from generation to generation in Vietnam, which relied on upholding, and sometimes gatekeeping, close therapy communities. This does not mean, however, that traditional medical knowledge is being 'democratised' as such. There is very little cross-pollination of knowledge sharing among the three sampled sites; people who participate in one site are not likely to also participate in others. Considering that each of these sites was organised around a different therapeutic regime (Southern medicine vs. 'Traditional' Northern medicine) with different philosophical outlooks (consumer advocacy vs. agenda-free knowledge sharing), the analysis seems to be suggesting that existing boundaries among different 'traditions' are being replicated online.

The lack of coordination across these sites might also have deeper roots in existing social mechanisms that maintain the propagation of traditional medical knowledge. Craig (2002) noted how the legacy of Vietnamese family health knowledge and practice, transmitted in its most durable forms through oral traditions and written recipes, is located within the household level where it is readily put to use. Since this locates the primary caregiving responsibility within the family unit rather than with professionals and institutions or online strangers, the logic that drives the propagation of non-biomedical knowledge is that of use-value: that people seek traditional medical knowledges in times of sickness and share them mostly in response to those in need in a transactional fashion. Unsolicited sharing of recipes and knowledges, when it happens, also seems to be grounded in collectively imagined boundaries between various undercurrents of TM. The lack of cross-pollination among different 'traditions' and consumer movements – Southern medicine vs. Northern medicine vs. Anti-biomedicine – seems to be replicating itself online, where people engage in rather insular and separate networks that map onto existing knowledge paradigms that are anchored in well-established everyday practices.

Despite the lack of interaction across different sites, network activities within the sites themselves are quite robust and resistant to change. All three networks exhibit 'small world' characteristics – which structurally enable quick and easy propagation of information. It is in this regard that Vietnamese non-biomedical networks resemble the characteristics of other networks on Facebook (Catanese et al., 2011; Caci et al., 2012, Wohlgemuth & Matache, 2012). This analysis contributes to the growing body of evidence of the ubiquity of small-world networks on Facebook, which could indicate that the affordances of Facebook as a platform might be shaping networks towards 'small-worldness'. If this is the case, then the growing popularity of self-contained community groups on Facebook might be fertile ground for resilient and durable discourse communities. Future research should look at the new temporalities that this mediated sociality is giving to the information and knowledge being propagated on social networking sites such as Facebook, especially with regard to rich and complex multimedia formats such as live-streaming videos and synchronous viewing of prerecorded videos.

With the LDA method, 49 unique topics were identified and qualitatively labelled. The significant number of overlapping topics found within the dataset is reflective of both the nature of social networking behaviour and the way TM is communicated in Vietnamese. Reposting popular and interesting content found elsewhere is common behaviour on social media, making frequently recurring content characteristic of social media data. LDA modelling picked up this pattern in the dataset.

Overall, the topics identified through LDA can fit under eight broad themes: managing health and illnesses (topics 2–12, 14–17, 19–20, 24–25, 28–33, 35–40, 45–49), the institutionalization of TM (topic 1), origins and legitimacy (topic 23), sales (topics 3, 7, 26, 27), lifestyle (topics 6, 21, 22, 34), religion and philanthropy (topic 4), negative aspects to TM (topics 13, 18, 42), and TM and overseas Vietnamese (topics 14, 16, 44). Among these broad themes, the last three themes are probably the most interesting. A close reading of posts containing the theme of religion and philanthropy reveals that Buddhist temples remain an important locus through which people of disadvantaged socioeconomic backgrounds in Vietnam seek and receive healthcare. TM, usually in the form of raw ingredients, is also frequently distributed for free by monks who practice medicine through Buddhist temples. This is an interesting finding, as it is pointing to the informal yet significant role religious institutions continue to play, especially in a secular, post-socialist society such as Vietnam. One of the earliest extant Vietnamese medical texts, 'Miraculous Drugs of the South' (*Nam Dược Thần Hiệu*), for example, was written by the Vietnamese Buddhist monk-physician Tuệ Tĩnh (ea. 1330-ca. 1389). For many centuries in Vietnam and East and Southeast Asia more generally, it was common for Buddhist monks and nuns to work as healers; Buddhist contexts have continued to be the most important loci for the cross-cultural exchange of diverse currents of medicine ideas and practices concerning illness and healing (Thompson, 2017). Local traditions of Buddhist medicine represent unique hybrid combinations of cross-culturally transmitted and indigenous knowledge. In addition to the transformations happening to Buddhist medicine by means of interactions with Western colonialism, scientific ideas, and new biomedical technologies, the internet and its social media platforms are the latest actors to contribute to the evolution and persistence of these non-biomedical modalities.

Critical discussions against TM are also present on these networks. There appear to be negotiations of what constitutes legitimate uses of medicinal plants, and indeed what counts as 'medicine' through these critical discussions. For example, in one discussion, speculations on the medicinal properties of shrimp paste – a Southeast Asian fermented condiment – were criticised as nonsensical and labelled as 'country bumpkin' thinking. 'Food as medicine' has long been a prominent characteristic in the East and Southeast Asian systems of medical thought, where local food cultures are inseparable from traditional therapeutic systems. The perceived multiple functions of edible plants and local food, however, are not immutable; as the above example shows, the medicinal functions of local food are subject to ongoing negotiation and reinterpretations as understandings about nutrition and health evolve. Future research could look into the ways in which living discourses surrounding policing and adjudicating the boundaries between food and medicine intersect with processes of urbanization and modernization, as well as how the changing distinction between functional foods and food medicines is being played out on social media.

Finally, it appears that the discourses conducted on and through these sites are transnational in nature. *Việt Kiều*, or overseas Vietnamese, are present in these online discussions and facilitate discussion topics that are distinct from those who reside inside Vietnam. These discussions involve, among others, requests for and provision of referrals to unlicensed traditional medical practices outside of Vietnam and transnational trading of herbal ingredients through informal means. It is estimated that there are around 4.5 million Vietnamese living overseas, contributing USD15.9 billion to the Vietnamese economy annually in remittances (Minh Huy, 2018). The majority of the Vietnamese diaspora left Vietnam as political and economic refugees at the end of the Vietnam War in 1975; almost half of overseas Vietnamese reside in the United States, and the majority of *Việt Kiều* live in other industrialised countries such as Japan, France, Australia, and Canada. Given that non-biomedical therapies are much more marginalised and stigmatised in these societies, future research could look at the ways in which diasporic communities navigate, with or without success, the healthcare systems of host states while forging and maintaining links with the 'homeland' through participating in networked propagation of traditional knowledges. Beyond issues concerning the navigation of biomedical health systems, issues with transnational belonging and emergent hybrid narratives about health and illness may also manifest themselves in novel ways through these networks. Furthermore, these network connections have the potential to materialise through the increasingly dense networks of transnational mobility. In Chapters 4 and 5, we will discuss how these results help inform the design of digital ethnographic methods to study emergent polymedia practices of diasporic actors as they find ways to manage everyday experiences of illness with digital technologies.

Note

1 For a concise review of this scandal in relation to the Graph API, see Albright (2018).

References

Albright, J. (2018, March 20). *The graph API: Key points in the Facebook and Cambridge analytica debacle*. Medium. https://medium.com/tow-center/the-graph-api-key-points-in-the-facebook-and-cambridge-analytica-debacle-b69fe692d747

Arun, R., Suresh, V., Madhavan, C. V., & Murthy, M. N. (2010). On finding the natural number of topics with latent dirichlet allocation: Some observations. In *Pacific-Asia conference on knowledge discovery and data mining* (pp. 391–402). Springer.

Baccianella, S., Esuli, A., & Sebastiani, F. (2010). Sentiwordnet 3.0: An enhanced lexical resource for sentiment analysis and opinion mining. In *Proceedings of the seventh international conference on language resources and evaluation (LREC'10)*. European Language Resources Association (ELRA).

Baldwin, T. (2012, November). Social media: friend or foe of natural language processing? In Proceedings of the 26th Pacific Asia Conference on Language, Information, and Computation (pp. 58–59).

Bassett, D. S., & Bullmore, E. T. (2017). Small-world brain networks revisited. *The Neuroscientist*, 23(5), 499–516. 10.1177/1073858416667720.

Blank, G. (2017). The digital divide among Twitter users and its implications for social research. *Social Science Computer Review*, *35*(6), 679–697. 10.1177/0894439316671698.

Blei, D. M. (2012). Surveying a suite of algorithms that offer a solution to managing large document archives. *Communication of the ACM*, *55*(4), 77–84.

Blei, D. M., & Lafferty, J. D. (2006). Dynamic topic models. In *Proceedings of the 23rd international conference on machine learning* (pp. 113–120). Association for Computing Machinery.

Blei, D. M., & Lafferty, J. D. (2007). A correlated topic model of science. *The Annals of Applied Statistics*, *1*(1), 17–35. 10.1214/07-AOAS114.

Blei, D. M., Griffiths, T. L., & Jordan, M. I. (2010). The nested chinese restaurant process and bayesian nonparametric inference of topic hierarchies. *Journal of the ACM (JACM)*, *57*(2), 1–30. 10.1145/1667053.1667056.

Blei, D. M., Ng, A. Y., & Jordan, M. I. (2003). Latent Dirichlet allocation. *Journal of Machine Learning Research*, *3*, 993–1022.

Blei, D. M., & Mcauliffe, J. (2007). Supervised topic models. In *Neural information processing systems proceedings* (pp. 121–128).

Bradley, M. M., & Lang, P. J. (1999). *Affective norms for English words (ANEW): Instruction manual and affective ratings* (Vol. 30, No. 1, pp. 25–36). Technical report C-1, The Center for Research in Psychophysiology, University of Florida.

Caci, B., Cardaci, M., & Tabacchi, M. E. (2012). Facebook as a small world: a topological hypothesis. *Social Network Analysis and Mining*, *2*(2), 163–167. 10.1007/s13278-011-0042-8.

Cao, J., Xia, T., Li, J., Zhang, Y., & Tang, S. (2009). A density-based method for adaptive LDA model selection. *Neurocomputing*, *72*(7–9), 1775–1781. 10.1016/j.neucom.2008.06.011.

Carolan, B. V. (2016). *Social network analysis*. Oxford Bibliographies in Education. 10.1093/OBO/9780199756810-0167.

Catanese, S. A., De Meo, P., Ferrara, E., Fiumara, G., & Provetti, A. (2011). Crawling facebook for social network analysis purposes. In *Proceedings of the international conference on web intelligence, mining and semantics* (pp. 1–8).

Compton, P., & Jansen, R. (1990). A philosophical basis for knowledge acquisition. *Knowledge Acquisition*, *2*(3), 241–257.

Craig, D. (2002). *Familiar medicine: Everyday health knowledge and practice in today's Vietnam*. University of Hawaii Press.

Csardi, G., & Nepusz, T. (2006). The igraph software package for complex network research. *InterJournal, Complex Systems*, *1695*(5), 1–9.

Dashtipour, K., Poria, S., Hussain, A., Cambria, E., Hawalah, A. Y., Gelbukh, A., & Zhou, Q. (2016). Multilingual sentiment analysis: state of the art and independent comparison of techniques. *Cognitive Computation*, *8*(4), 757–771. 10.1007/s12559-016-9415-7.

Denecke, K. (2008). Using sentiwordnet for multilingual sentiment analysis. In *2008 IEEE 24th international conference on data engineering workshop* (pp. 507–512). IEEE.

DiMaggio, P. (2015). Adapting computational text analysis to social science (and vice versa). *Big Data & Society*, *2*(2), 1–5. 10.1177/2053951715602908.

Dixon, S. (2022). *Countries with the most Twitter users 2022*. Statista. https://www.statista.com/statistics/242606/number-of-active-twitter-users-in-selected-countries/.

Epskamp, S., Costantini, G., Haslbeck, J., Cramer, A. O., Waldorp, L. J., Schmittmann, V. D., & Borsboom, D. (2019). *Package 'qgraph'*. R-Project. https://cran.r-project.org/web/packages/qgraph/qgraph.pdf.

Evans, M. S. (2014). A computational approach to qualitative analysis in large textual datasets. *PloS One*, *9*(2), e87908. 10.1371/journal.pone.0087908.

Farzindar, A. A., & Inkpen, D. (2020). *Natural language processing for social media* (3rd ed.). Springer Nature.

Fine, A., Crutchley, P., Blase, J., Carroll, J., & Coppersmith, G. (2020). Assessing population-level symptoms of anxiety, depression, and suicide risk in real time using NLP applied to social media data. In *Proceedings of the fourth workshop on natural language processing and computational social science* (pp. 50–54).

Frey, B. (2018). *The SAGE encyclopedia of educational research, measurement, and evaluation* (Vols. 1–4). SAGE Publications, Inc. 10.4135/9781506326139.

Gerlach, M., Peixoto, T. P., & Altmann, E. G. (2018). A network approach to topic models. *Science Advances*, 4(7), eaaq1360. 10.1126/sciadv.aaq1360.

Graham, T. & Ackland, R. (2016). *SocialMediaLab: Tools for collecting social media data and generating networks for analysis*. CRAN (The Comprehensive R Archive Network). https://rdrr.io/cran/SocialMediaLab/.

Griffiths, T. L., & Steyvers, M. (2004). Finding scientific topics. *Proceedings of the National Academy of Sciences*, 101(suppl 1), 5228–5235. 10.1073/pnas.0307752101.

Grün, B., & Hornik, K. (2011). Topicmodels: An R package for fitting topic models. *Journal of Statistical Software*, 40(13), 1–30.

Habernal, I., Ptáček, T., & Steinberger, J. (2014). Supervised sentiment analysis in Czech social media. *Information Processing & Management*, 50(5), 693–707. 10.1016/j.ipm.2014.05.001.

Hanneman, R. A., & Riddle, M. (2016). Concepts and measures for basic network analysis. In J Scott, & P J Carrington (Eds.), *The SAGE handbook of social network analysis*. Sage.

Hether, H. J., Murphy, S. T., & Valente, T. W. (2016). A social network analysis of supportive interactions on prenatal sites. *Digital health*, 2, 1–12. 10.1177/2055207616628700.

Hoang, T. A. (2015). Modeling user interest and community interest in microbloggings: An integrated approach. In *Pacific-Asia conference on knowledge discovery and data mining* (pp. 708–721). Springer.

Hou-Liu, J. (2018). Benchmarking and improving recovery of number of topics in latent dirichlet allocation models. *viXra*. https://pdfs.semanticscholar.org/2175/aa77463e23da96281cc2fb5125e0b9de3bbd.pdf.

Humphries, M. D., & Gurney, K. (2008). Network 'small-world-ness': a quantitative method for determining canonical network equivalence. *PLoS One*, 3(4), e0002051. 10.1371/journal.pone.0002051.

Jones, K. S. (1994). Natural language processing: a historical review. In A Zampolli, N Calzolari, & M Palmer (Eds.), *Current issues in computational linguistics: in honour of Don Walker* (pp. 3–16).

Kornai, A. (2013). Digital language death. *PLoS One*, 8(10), e77056. 10.1371/journal.pone.0077056.

Le, H. P., Nguyen, T. M. H., Roussanaly, A., & Ho, T. V. (2008). A hybrid approach to word segmentation of Vietnamese texts. In *Proceedings of the 2nd international conference on language and automata theory and applications* (pp. 240–249).

Li, W., & McCallum, A. (2006). Pachinko allocation: DAG-structured mixture models of topic correlations. In *Proceedings of the 23rd international conference on Machine learning* (pp. 577–584).

Maier, D., Waldherr, A., Miltner, P., Wiedemann, G., Niekler, A., Keinert, A., Pfetscha, B., Heyerc, G., Reberd, U., Häusslerd, T., Schmid-Petrie, H. & Adam, S. (2018). Applying LDA topic modeling in communication research: Toward a valid and reliable methodology. *Communication Methods and Measures*, 12(2–3), 93–118. 10.1080/19312458.2018.1430754.

Marin, A., & Wellman, B. (2011). Social network analysis: An introduction. In J Scott, & P J Carrington (Eds.), *The SAGE handbook of social network analysis* (pp. 11–25). Sage.

Mimno, D., Wallach, H., Talley, E., Leenders, M., & McCallum, A. (2011). Optimizing semantic coherence in topic models. In *Proceedings of the 2011 conference on empirical methods in natural language processing* (pp. 262–272).

Minh Huy (2018, December 28). *Overseas remittances to Vietnam continue increasing*. Sai Gon Giai Phong News Online. https://m.sggpnews.org.vn/business/overseas-remittances-to-vietnam-continue-increasing-79438.html.

Monnais, L., Thompson, C. M., & Wahlberg, A. (Eds.). (2011). *Southern medicine for southern people: Vietnamese medicine in the making*. Cambridge Scholars Publishing, UK.

Mullen, T., & Collier, N. (2004). Sentiment analysis using support vector machines with diverse information sources. In *Proceedings of the 2004 conference on empirical methods in natural language processing* (pp. 412–418).

Newman, M. E. (2009). Random graphs with clustering. *Physical Review Letters, 103*(5), 058701. 10.1103/PhysRevLett.103.058701.

Newman, M. E., Moore, C., & Watts, D. J. (2000). Mean-field solution of the small-world network model. *Physical Review Letters, 84*(14), 3201. 10.1103/PhysRevLett.84.3201.

Nguyen, D. (2021a). Dropping in, helping out: Social support and weak ties on traditional medicine social networking sites. *Howard Journal of Communications, 32*(3), 235–252. 10.1080/10646175.2021.1878478.

Nguyen, D. (2021b). The network life of non-biomedical knowledge: Mapping vietnamese traditional medicine discourses on facebook. *Journal of Digital Social Research, 3*(2), 10–43. 10.33621/jdsr.v3i2.82.

Nguyen, D. Q., Nguyen, D. Q., Pham, S. B., Nguyen, P. T., & Nguyen, M. L. (2014). From treebank conversion to automatic dependency parsing for Vietnamese. In *International conference on applications of natural language to data bases/information systems* (pp. 196–207). Springer.

Nguyen, D. Q., Nguyen, D. Q., Vu, T., Dras, M., & Johnson, M. (2018). A fast and accurate Vietnamese word segmenter. In *Proceedings of the eleventh international conference on language resources and evaluation (LREC 2018)* (pp. 2582–2587).

Nikita, M. (2016). Package 'ldatuning': Tuning of the latent dirichlet allocation models parameters. r package version 1.0.0. https://CRAN.Rproject.org/package=ldatuning.

Opsahl, T., Vernet, A., Alnuaimi, T., & George, G. (2017). Revisiting the small-world phenomenon: efficiency variation and classification of small-world networks. *Organizational Research Methods, 20*(1), 149–173. 10.1177/1094428116675032.

Ortigosa, A., Martín, J. M., & Carro, R. M. (2014). Sentiment analysis in Facebook and its application to e-learning. Computers in Human Behavior, 31, 527–54110.1016/j.chb.2013.05.024.

Otter, D. W., Medina, J. R., & Kalita, J. K. (2021). A survey of the usages of deep learning for natural language processing. *IEEE Transactions on Neural Networks and Learning Systems, 32*(2), 604–624. 10.1109/tnnls.2020.2979670.

Pang, B., & Lee, L. (2008). Opinion mining and sentiment analysis. *Foundations and Trends® in Information Retrieval, 2*(1–2), 1–135. 10.1561/1500000011.

Petrov, S. (2016, May 12). *Announcing syntaxnet: The world's most accurate parser goes open source*. Google Research. https://ai.googleblog.com/2016/05/announcing-syntaxnet-worlds-most.html.

Seki, Y., Ku, L. W., Sun, L., Chen, H. H., & Kando, N. (2010). Overview of multilingual opinion analysis task at NTCIR-8. In *Proceedings of the 8th NTCIR workshop meeting on evaluation of information access technologies* (pp. 209–220).

Sievert, C., & Shirley, K. (2014). LDAvis: A method for visualizing and interpreting topics. In *Proceedings of the workshop on interactive language learning, visualization, and interfaces* (pp. 63–70).

Strapparava, C., & Valitutti, A. (2004). Wordnet affect: an affective extension of wordnet. In *LREC* (Vol. 4, No. 1083–1086, p. 40).

Tan, S., & Zhang, J. (2008). An empirical study of sentiment analysis for Chinese documents. *Expert Systems with Applications, 34*(4), 2622–2629. 10.1016/j.eswa.2007.05.028.

Tausczik, Y. R., & Pennebaker, J. W. (2010). The psychological meaning of words: LIWC and computerized text analysis methods. *Journal of Language and Social Psychology, 29*(1), 24–54. 10.1177/0261927X09351676.

Telesford, Q. K., Joyce, K. E., Hayasaka, S., Burdette, J. H., & Laurienti, P. J. (2011). The ubiquity of small-world networks. *Brain Connectivity, 1*(5), 367–375. 10.1089/brain.2011.0038.

Thang, D. Q., Phuong, L. H., Huyen, N. T. M., Tu, N. C., Rossignol, M., & Luong, V. X. (2008). Word segmentation of Vietnamese texts: a comparison of approaches. In *Proceedings of the 6th international conference on language resources and evaluation* (pp. 1933–1936).

Thelwall, M., Buckley, K., & Paltoglou, G. (2011). Sentiment in Twitter events. *Journal of the American Society for Information Science and Technology*, 62(2), 406–418. 10.1002/asi.21462.

Thompson, C. (2017). 57. Selections from miraculous drugs of the south, by the Vietnamese Buddhist Monk-Physician TueTinh. In *Buddhism and medicine* (pp. 561–568). Columbia University Press. 10.7312/salg17994-059.

Unicode. (2015, December 16). *Unicode launches adopt-a-character campaign to support the world's "digitally disadvantaged" living languages*. The Unicode Blog. http://blog.unicode.org/2015/12/unicode-launches-adopt-character.html.

Uzzi, B., & Spiro, J. (2005). Collaboration and creativity: The small world problem. *American Journal of Sociology*, 111(2), 447–504. 10.1086/432782.

Uzzi, B., Amaral, L. A., & Reed-Tsochas, F. (2007). Small-world networks and management science research: A review. *European Management Review*, 4(2), 77–91. 10.1057/palgrave.emr.1500078.

van Dijck, J., & Poell, T. (2013). Understanding social media logic. *Media and Communication*, 1(1), 2–14. 10.17645/mac.v1i1.70.

Vu, T., Nguyen, D. Q., Nguyen, D. Q., Dras, M., & Johnson, M. (2018). *VnCoreNLP: A Vietnamese natural language processing toolkit*. arXiv. 10.48550/arXiv.1801.01331.

Wali, E., Chen, Y., Mahoney, C., Middleton, T., Babaeianjelodar, M., Njie, M., & Matthews, J. N. (2020). Is machine learning speaking my language? A critical look at the NLP-pipeline across 8 human languages. *37th International conference on machine learning*.

Wang, H., Can, D., Kazemzadeh, A., Bar, F., & Narayanan, S. (2012). A system for real-time twitter sentiment analysis of 2012 us presidential election cycle. In *Proceedings of the ACL 2012 system demonstrations* (pp. 115–120).

Wang, X., & McCallum, A. (2006). Topics over time: a non-markov continuous-time model of topical trends. In *Proceedings of the 12th ACM SIGKDD international conference on Knowledge discovery and data mining* (pp. 424–433).

Watts, D. J. (1999). Networks, dynamics, and the small-world phenomenon. *American Journal of Sociology*, 105(2), 493–527. 10.1086/210318.

Watts, D. J., & Strogatz, S. H. (1998). Collective dynamics of 'small world' networks. *Nature*, 393(6684), 440–442.

Wohlgemuth, J., & Matache, M. T. (2012). *Small world properties of Facebook group networks* [Master's thesis, University of Nebraska at Omaha]. ProQuest Dissertations Publishing.

Zaugg, I. A., Hossain, A., & Molloy, B. (2022). Digitally-disadvantaged languages. *Internet Policy Review*, 11(2). 10.14763/2022.2.1654.

PART II

Qualitative and technological approaches

4

DIGITAL DIASPORAS AS ETHNOGRAPHIC SITES

Introduction

Where are digital diasporas and how do we go about assembling them? Implicit in the language we use to talk about diasporas is the contention that diasporas are spatial configurations through which social interactions occur and thus can be observed. An ethnographer of digital diasporas would presumably 'hang out' in a diasporic space of their choosing using techniques and approaches not too dissimilar to those of a traditional ethnographer: they would watch the events that unfold over time, listen to what is said, ask questions, write up copious amounts of fieldnotes, and reflect on their role throughout the research process. But given that digital diasporas are mediated spaces which elude direct physical presence, how do we know if/when we are in the right place, or indeed when we are in a place at all? Where does a diasporic site begin and when does it end?

This chapter deals with the question of space in ethnographic research of digital diasporas. There is a double sense of place in the notion of digital diaspora – that of the spatial digital and that of the spatial diaspora. While discussions on the pitfalls of spatial conceptualisations of the internet (Graham, 2013; Champion, 2021) and of diasporas as contingently spatialised and as spatial containers (Axel, 2004; Brubaker, 2005) are well-rehearsed, little has been said on the implications of mobilising these two spatial concepts alongside each other to describe the emergent, relational, and mediatised practices of diasporic actors. Little has been said, also, on the practical implications of this formulation for researchers interested in studying diasporic lives at the intersection of technology and culture. To what extent is the notion of digital diaspora useful to researchers, and how should it be mobilised to better understand the wide-ranging impact that digital technologies are having on the way diasporic lives are enacted across (and within) national borders? Insofar as experiences of digital space are often invoked among diasporic actors themselves, how should researchers make sense of these testimonials, and how do these spatial conceptions help refresh approaches to topics of interest to diaspora research?

DOI: 10.4324/9781003336556-7

The chapter starts by discussing digital diasporas as complex arrangements of difference that require new vocabulary beyond that of 'culture' and 'society' as autonomous entities. The chapter encourages researchers to complicate digital diasporas beyond obvious, ready-made, and technically circumscribed sites of diasporic activities (such as diasporic forums, social media groups, or microblogging discourse communities), and instead understand digital diasporas as ongoing enactments of shared identity, common meaning, and imagined sovereignty that is nevertheless always subject to transcension. Following such an approach requires researchers to always be prepared to attend to events that may disrupt and unfasten any preconceived notion about social relations and be prepared to be taken by the actors involved in this process to unfamiliar territories. The chapter also discusses the implications of shifting diasporic sociality as it is co-produced alongside rapid changes in digital technologies, focusing on how the task of determining, locating, and assembling digital diasporic sites can be informed by the quantitative methods discussed in Part I. The chapter concludes with a case study on conceptions of mediated space among Vietnamese diasporas and makes the case for attending to space as an explicit conceptual category in digital diaspora studies – not only to better understand how diasporic actors experience digital space and how this experience shapes and transforms their everyday lives, but also to sharpen our oftentimes unarticulated theory of space in ethnographic practices.

Give me the internet and I'll build a diaspora

Digital technologies are widely acknowledged to have had a significant impact on the performance and formation of digital diasporas across various literatures, from media studies to diaspora studies and migration studies (Aziz, 2022; Brühne & Kuhlmann, 2022; Ponzanesi, 2020). The exact nature of this impact is a matter of ongoing empirical inquiry which varies across time and context; Nedelcu (2019), however, argued in general that technologies have both conservative and transformative effects on diasporas. The availability of digital technologies at low cost, together with increasingly widespread internet adoption, has allowed diasporic groups to deepen the quality of ties being forged and maintained with their country of origin. This deepening of quality is thanks to increases in frequency and intensity of diasporic communication, made possible as a result of both fundamental changes in the structure of mediated communication and thus the communication environment at large, as well as the emergence of new possibilities of diaspora enactment thanks to the media-rich characteristics of the internet. Nedelcu (2019) argued that transformations of diasporic functioning can be mapped against three main aspects: the actualisation of the homeland as memory on a day-to-day basis, the replacement of diaspora as 'non-lieu' by a sense of shared virtual place, and the agency capability of acting transnationally in real-time. In this light, digital diaspora is not only a new enactment of diaspora (which does not replace or supersede the old notion of diaspora as population dispersion) but also becomes an expression of the cosmopolitan condition. Digital formations facilitate and transform the very possibilities for diasporic affiliation; digital diaspora as space is therefore dynamic, processual, relational, and contingent.

While evocative and useful in its explanatory power, this conceptualisation of digital diaspora is necessarily porous in its boundaries. If digital diaspora is contingent and processual, what conceptual lineage does it share with traditional formulations of diaspora as bounded by long histories of trauma and loss, of dispersal and expropriation, of fixed relations to nationhood, citizenship, and metropolitan assimilation? Candidatu and Ponzanesi (2022) insist that in formulating digital diasporas alongside constantly changing digital media affordances, we are in fact attending to new configurations that enable, sustain and multiply diasporic encounters through social media platforms, digital devices, and infrastructures. Digital diaspora is articulated as fluid and relational because what it means to enact diasporas has changed by means of engagement with technology: processes of becoming diasporan are increasingly subject to ever more complex articulations and negotiations, and with them come new configurations of participation and identification that – while maintaining a well-established orientation towards notions of homeland, origin culture, and senses of belonging – are increasingly enmeshed in other networks of practice that make the diasporic experience increasingly elastic. As a heuristic tool that opens up epistemological and methodological possibilities for empirical research, digital diaspora is 'mutually constituted here and there, through bodies and data, across borders and networks, online and offline, by users and platforms, through material, symbolic, and emotional practices that are all reflective of intersecting power relations' (Candidatu et al., 2019, p. 34). For digital diaspora to remain analytically useful, it, therefore, has to be understood in relational terms, always as the unfolding of new identities and virtual communities that are informed by new forms of communication that recalibrate and intensify patterns of mobility and hybridity (Ponzanesi, 2020). This formulation also helps researchers avoid 'methodological nationalism' (Wimmer & Schiller, 2002) – wherein the trappings of the specificity of groups, ethnicities, and national contexts could foreclose a researcher's endeavour with a tunnel vision that could severely limit the scope, usefulness, and truthfulness of a research project.

To the extent that digital diaspora is a space, it is not one that exists prior to the researcher's entry. Nor does it strictly map onto the geographies of sovereign nation-states, with whom the strong ties maintained by diasporic actors form the very basis of digital diaspora. Digital diaspora is internet-specific, network-oriented, and embedded in wider social practices (Ponzanesi, 2020); it refers to hybrid spaces of belonging through efforts to maintain and reproduce shared cultural norms outside an at once actual and imagined homeland. Digital diaspora is a site of connectivity, where people gather on the basis of shared places of origin and of roots that precede the creation of digital diasporic spaces, and yet it is necessarily a site of multiple coexisting histories and geographies. Ponzanesi (2020) calls on the metaphor of the constellation – one that shares strong relational evocations with the network metaphor – and urges us to rethink digital diasporas as 'constellations within the digital firmament, rather than a technological matrix of connectivity' (p. 987). This conceptualisation also allows researchers to unfix and de-essentialise the ethnographic site as pre-made and readily available – and instead make the conceptual shift to ground their ethnographic practice in sites and places that are constantly in flux, mutations, and renegotiations. In looking for and assembling digital diasporas, researchers need to first foreground different experiences of locality, mobility, and digitality in correspondence with a mix of research techniques that allow them to identify traces of digital connectivity, exchange, and interaction – which in themselves can be fleeting and self-contained. These traces, once collected and analysed, allow researchers to begin sketching forward – deciding which interactions are more interesting, insightful, or carrying more potential. These analytical choices *make* the sites of digital diaspora; whether certain constructions of digital diaspora are more useful, compelling, or revelatory than others, therefore, depends on the methodological decisions that give shape to these sites.

Defining the ethnographic fieldsite as a heterogeneous network, Burrell (2017) contends with a relational conception of the ethnographic site as made up of constant movement, of overlapping and intertwined social terrain that transcends the old online-offline divide. Seeing the ethnographic site as a network helps researchers overcome the constraints of ethnography as originating in the anthropological traditions of visiting far-flung places to study wholes-of-culture, which has faithfully served as a strategy to unseat presumptions of the universal or biological basis of social practices through the articulation of cultural differences. But as this strategy becomes less appealing as the world becomes more and more interconnected – making the culture containment thesis increasingly less and less accurate – the practice of ethnography has evolved to orient itself around investigating the embedding of cultural worlds in larger impersonal systems of political economy. Such an evolution has also been fuelled by theoretical developments from within the field which contested the containment of culture, alongside the more open-ended practices happening from outside of anthropology (Burrell, 2017). As a result, fieldsite selection becomes an ongoing task, continually drawn and redrawn throughout the process of data gathering – rather than being decided once and for all in the early stages of the research.

Because the ethnographic site is no longer predetermined and fixed, the practical issue facing researchers has become that of foregrounding, where researchers are tasked with making coherent and distinct spaces of practice that are deeply embedded in the complex terrains of social lives. To this end, Burrell (2017) outlined six practical steps to help researchers locate and construct their fieldsite, drawing on her own experience conducting fieldwork in and around internet cafes in Ghana: seek entry points rather than sites, consider multiple types of networks, follow – but also intercept, attend to what is indexed in interviews, incorporate unhabitable spaces, and know when and where to stop. For researchers of digital diaspora, these steps could be contextualised by firstly foregrounding the phenomenon of digital diaspora as necessarily, but not exclusively, involving the interaction and exchange of diasporic actors who come together on the basis of shared identity and common meaning – practices that are oriented around notions of homeland, origin culture, and senses of belonging. From these broad contours, researchers can work on starting to engage reflexively with discourses that are of interest to diasporic lives on the internet as an entry point, and from there begin the tracing of their fieldsite by following diasporic actors and their practices, which would likely span across issues, offline/online divide, and geographical contexts. Researchers can familiarise themselves with discourses of interest to diasporic lives through various means – either from existing empirical literature, lived experience, and through competency in specific diasporic languages. In diaspora research, language competency is a strong asset (if not a prerequisite) to developing the cultural competency that allows researchers to immerse themselves into, and confidently traverse, the dynamics of diasporic worlds. Foregrounding as preceding the assemblage of a fieldsite allows researchers to be more precise in their empirical tracing of the target social phenomenon, since this is independent of the superimposed boundaries of the nation, the city, and so on – which do not necessarily map onto digital diasporic activities. Foregrounding sharpens the construction of the ethnographic site into a terrain that more closely tracks the digital diaspora under study – rather than assuming a naturally occurring, ready-made diasporic field of events within which researchers can dwell.

Once an entry point is established, researchers should explicitly seek out heterogeneity in their fieldwork. Burrell (2017) suggested that researchers attend to a variety of technological infrastructures such as phone networks, other telecommunications networks (such as the Internet), transportation networks (such as airlines), road networks, and social networks – and make explicit arguments about the constraining and enabling effects that these infrastructures have on the construction of the ethnographic site. In this sense, digital diasporas are never detached from the flows that animate other modes of connections which are entwined in diasporic lives: those pertaining to the state, market, and media. Following this, researchers of digital diaspora should move to determine a base – a point of intersection – from which to not only observe but also intercept with flows of conversations and circulations of materiality. In my fieldwork with Vietnamese diasporas, my main base was a *diện chẩn* clinic in Ho Chi Minh City where both locals and overseas Vietnamese would frequent to seek treatment. Even though my fieldwork eventually extended to various sites in the US (in the sense that I had the opportunity to really 'follow the actors'), the awareness of the multisited context of my main fieldsite allowed me to map out the flows of capital and health knowledge that travel into and out of this base, which subsequently allowed me to sketch out the common temporal structures of boredom and pain that allow these flows to converge the way they do (more on this in Chapter 5 case study). Burrell (2017)'s advice to listen to what is said in interviews as a way to augment the single site with awareness of other sites in the background is also crucial in helping researchers construct a fuller sense of place; interviewees, once comfortable about sharing their life stories with researchers, will usually bring up sites such as school, workplace, hospital, restaurant, a friend's house, a neighbourhood, and so on – all of which should be indexed against the main field base as potential locales to visit and as part of the spatial makeup of the ethnographic site.

The final steps – which are to incorporate uninhabitable spaces and to stop drawing the fieldsite where appropriate – encourage researchers both to attend to in situ conceptions of space as recounted by research participants and to be attuned to when a space has become saturated with meaning and thus no longer requires the researcher's presence. Burrell (2017), reflecting on her fieldwork in Accra, realised that many Ghanaian internet users conceived of chat rooms, dating websites, and other online spaces designed for mixing and mingling as providing access to philanthropists, potential business partners, and wealthy older people – a conception that which fails to map against their encounters with teenagers and twenty-somethings in these spaces. Uninhabitable spaces are spaces of possibility, where imaginations and projections of hope, longing, and even magic often play out in the most surprising ways – as we will see in the case study later in this chapter. Knowing when to stop drawing the field and start writing up research findings often comes naturally to the researcher after a period of time, as themes start to emerge, and information starts to repeat itself over time and across different research participants. This does not mean that digital diaspora ends where researchers decide to stop; it just means that to continue assembling and constructing digital diaspora, researchers will need to go back to where they started and embark in different directions, chase after different leads, and follow different actors. As ethnography evolves to orient itself around spaces as open-ended networks and assemblages and away from the static (and outdated) notion of whole-of-culture, the question of completeness becomes unproblematic: researchers stop their fieldwork when they must. Because researchers play a constitutive role in the construction of the ethnographic site, the key issue becomes not that of completeness, but rather of reflexivity, systematicity, and conceptual precision. The next section discusses the need to incorporate methods and approaches from different research traditions to help researchers more creatively and effectively construct digital diasporas as they address these quality benchmarks.

Diasporic sociality and digital technologies: a call for mixing methods

Besides drawing from literature, lived experiences, and language competency to locate entry points into ethnographic sites, researchers can also draw on a variety of computational techniques to help assist this task. Beaulieu (2017) made a useful distinction between mediated ethnography and computational ethnography to highlight how different equipment – both in the literal and epistemological senses – constitute different modes of problematisation in doing ethnography. Mediated ethnography, she argued, is concerned primarily with the implications of recognising mediation as a prominent part of fieldwork and field relations. In other words, mediated ethnography recognises systematic engagement with digital technologies as a means through which fieldwork might occur, where well-established ethnographic techniques might be reproduced, adapted, and improvised for the mediated environment. Mediated ethnography as a practice is concerned with the experience of 'being there' in spite of, through, and thanks to mediation – so that the ethnographer might gain first-hand experience about the practice under study. Tensions between the digital and the analogue animate much of mediated ethnography, where researchers are attuned to the ways in which ethnography practices might be changed, augmented, and challenged by social relations that unfold over mediated environments.

Computational ethnography, on the other hand, is centred on the informational dimension of internet settings, big data, and digital tools (Beaulieu, 2017). For Beaulieu, the productive tension that animates computational ethnography is not with the analogue, but rather with the unfolding of events – the narrative – which is central to ethnographic practice and accounts. Citing the transition of the web from the notion of hyperlinks being 'authored' by web designers to the automation of ethnographic spaces by algorithms, profiling, and scripts, Beaulieu describes computational ethnography as a practice that foregrounds computational aspects of code and of networks as objects of concern. Computational ethnography is particularly attuned to the ways that suites of technologies – rather than a particular platform for forums – constitute the fieldsite, giving it meaning and context. While Beaulieu (2017) saw computational means of accessing the site as somewhat restricted to data collection that is surreptitious and without explicit elicitation, where interactions and behaviours of interest are captured through automated means, I argue in my case study below that automated data collection informs the construction of ethnographic sites by producing the kind of knowledge specific to its own underlying epistemology. In other words, automated data collection as an exercise does not equate to getting access to the fieldsite; what it can do by virtue of generating its own knowledge is provide researchers with yet another source of information to consult when they deliberate on finding the appropriate entry point, which actors to follow, which leads to chase after, which connections and interactions to privilege in a complex network of relationships, and so on. Furthermore, while Beaulieu (2017) sees the analysis of automated data as part of ethnography's commitment to going beyond formal accounts – in the specific meaning that analysing what people do to augment the account of what they say they do enables researchers to get a more informed picture of what goes on in the field – I argue that working with computational methods actually is the formal approach that could complement the informal (open-ended, interpersonal, immersive) nature of ethnography practices.

What are some implications of mixing methods to locate ethnographic sites in this way? In the context of digital diaspora studies, a commitment to mixing methods and experimenting with bringing together different epistemologies allows researchers to learn the nuances of the digital as it is expressed and not expressed in everyday lives, and to move beyond an ethnonational framework in addressing the possibilities the digital media afford. Witteborn (2019), for example, reconsiders the explanatory power of the digital diaspora and proposes an understanding of the digital diaspora through the lens of loss in her fieldwork, urging researchers to shift their vision from the stuck-ness of ethnic and national sameness to the possibilities of 'diasporic solidarities beyond the ethnonational bond' (p. 184). In this way, digital diaspora as a phenomenon and as a conceptual framework can be mobilised to account for the different material and affective dimensions of mediated displacement and loss that happen as a result of mobility and migration. Elsewhere, Mitra (2008) argues that Indian call centre employees can be understood as a form of diaspora community, since they are located locally in their country of origin but have to adapt to the culture and customs of Western customers in their workplace. In subverting the physical location as the place of origin and the virtual location as the place of adoption for this community, Mitra (2008) makes the case that digital diaspora as a process of ongoing displacement and (dis)connection can be designed against the background of the connotations of diaspora. These are but among the many ways in which the spaces of digital diaspora can be constructed through reflexive experimentation with methods as epistemologies – in response to complex empirical realities and in accommodation with domain-specific theoretical frameworks.

Mixing methods is not the same as a methodological free-for-all, however. In attempting to bring together different methods so as to innovate ourselves out of disciplinary impasses and closer to the phenomenon we wish to understand, there are epistemological grounds on which researchers should tread carefully, or else they would risk incoherence and inconsistency in their research design. Ardèvol and Lanzeni (2017) insist that as researchers work with ethnographic methods, they should build their research design on three pillars: displacement, denaturalisation, and open theory. Displacement refers to the distinction between the self and the other – the experience of which motivates ethnography as a method to understand others and to grasp the diversity of life forms. This sense of displacement drives the researcher to try to experience the world *with* others as they do on their own terms; in order to do this, the researcher's ideas are always contrasted with the native concepts – a practice Ardèvol and Lanzeni (2017) refer to as denaturalisation. For them, ethnographic knowledge is the comprehension that comes out of denaturalisation; ethnographic knowledge usually demands adjustment and reformulation of the researcher's theories and conceptual frameworks, making ethnographic research necessarily open and nonlinear. This open theory nature requires researchers to work on shifting grounds, allowing pre-concepts and theories to remain open to change during fieldwork. In practice, much (if not all) of research is non-linear; even with the sequential workflows that are baked into the design of much of computational, quantitative methods, researchers often find themselves going back and forth between various steps of the research process to finetune their design, tweak parameters, ask better questions, change course, wrangle with data, abandon interpretations, do more fieldwork, redesign analysis schemes. What is specific to mixing methods with ethnography is the full embrace of this non-linearity not as a bug, but as a feature of the research process; not only are detours and dead ends acknowledged and welcomed in an ethnographic research mix, but they are also foregrounded, carefully documented, reflected upon, and incorporated in the presentation of research findings.

In the next section, I will present a case study on how spatial understandings of the internet (i.e., the internet is a 'space') enable the persistence of marginalised medical practices among the Vietnamese diaspora. This case study is an example of mixing computational analysis with ethnographic methods. The case study was motivated by the findings discussed in Chapter 3, where topic modelling results pointed me to issues with transnational belonging and emergent hybrid narratives about health and illness that are present within Vietnamese diasporic groups on Facebook. Following this important finding, which was unexpected in my original focus on traditional medicine groups on Facebook, I decided to expand my scope of research to include *diện chẩn* – an emergent Vietnamese non-biomedical therapy. By tracing different accounts of the internet as space among practitioners and followers of *diện chẩn*, I show how the logic of space circumscribes an alternative techno-social site for marginalised medical practices, transforms the private experience of being alone with technology into being-in-space, spiritualises the internet as a conduit of healing power, and mediates transnational health mobility among the Vietnamese diaspora. This case study highlights how a sharp focus on the ethnographic site as space could help researchers of digital diaspora refresh their approach to understanding and explaining phenomena that are otherwise 'blackboxed' as irrational or nonsensical. It also demonstrates how this focus could contribute to theoretical discussions on the spatial internet, and in so doing allowing researchers to contribute productively at the intersection of multiple research literatures.

Case study: conceptions of mediated space among Vietnamese diasporas

This section reports on the methods and key findings in Nguyen et al. (2021) as a case study for doing digital ethnography that transcends the online/offline dichotomy. The section introduces the domain-specific literature that the paper contributes to by briefly introducing its theoretical framework and its research questions before reporting in detail on the practical methodological choices made in its analysis. The case study demonstrates how enactments of digital diasporas are deeply material and sensorial, and that diasporic connections are never divorced from temporal structures and material conditions of the various localities they purport to transcend. To attend to these temporal and spatial configurations is to explicitly account for the materiality of devices, bodies, and context in ethnographic research design. The case study shows how *in vivo* conceptualisation of the internet as space offers a point of convergence against the bifurcation of information as abstract and technology as concrete. A spatial conceptualisation highlights the embeddedness of health knowledge on the internet and shows how techno-social interrelations produce different spaces of multiplicity which constitute a favourable milieu for medical practices outside of the biomedical institution to persist.

Background to case study

Diện chẩn (literally 'face diagnosis' – hereafter DC) is an emergent practice of therapy that offers new interpretations of Vietnamese and Chinese traditional medicine and new inventions of therapeutic tools. Behind this emergent Vietnamese non-biomedical practice, which has attracted an international following via the internet, is an obscure man named Bùi Quốc Châu, who claims to have single-handedly invented a new way to diagnose and treat most diseases through particular ways to massage the face. Particularly, different constellations of points on a human face, according to DC, are said to correspond to different organs in the human body. If these points are massaged correctly according to the method, the claim is that any and all diseases within the corresponding organs can be cured. For example, intense and frequent massaging of the mento labial sulcus (area between the lower lip and the chin) using a DC tool is claimed to cure uterine fibroids, as the area is believed to correspond to the uterus in the female body. DC claims to have gathered millions of followers in 35 different countries (Vu, 2020). While it is difficult to ascertain the veracity of this claim, our own documentation of this method shows that DC materials are available online in at least nine different languages (Vietnamese, Spanish, Polish, English, Italian, French, Portugese, German, Russian).

During the 1980s, Bùi Quốc Châu's claim of inventing a unique method to diagnose and treat diseases through facial massage gained attention from important politicians in the politburo. In 1988, he was sent to Cuba to promote 'Vietnamese medicine' and enhance his skills through 'scientisation,' which was a significant political-medical project in Cuba at that time (Sai Gon Giai Phong, 1988). However, his mission was unsuccessful, and in the early 1990s, when power shifted within the politburo, Diện chẩn fell out of favour. The DC research centre in Ho Chi Minh City was later confiscated by the government, and Bùi Quốc Châu's attempts to gain regulatory recognition from the Vietnamese government were unsuccessful. In 2003, an article in Lao Động newspaper highlighted the lack of scientific evidence for DC as a therapeutic method (VNExpress, 2003). In 2012, Sài Gòn Giải Phóng newspaper conducted a three-part investigation into DC, which concluded that not only is there no scientific basis for this practice, but it may also be illegal (Sai Gon Giai Phong, 2012a, b, c).

As such, the space for DC practice – social, political, and legal – has repeatedly been curtailed as post-war, post-socialist Vietnam comes to integrate certain non-biomedical practices into its mainstream biomedical healthcare system and parochialise others. This has led to a nested hierarchy of medical traditions in Vietnam and challenges for DC to either adapt or decline. However, the rise of digital technology and internet usage in

Vietnam has allowed for alternative spaces of practice to emerge, where different non-biomedical practices can coexist and compete for attention from users. Few studies have explored how these alternative spaces impact how people manage their health and illness experiences.

My methodological interest in reporting on this case study is threefold. Firstly, this case study demonstrates how when researchers render their spatial conception of the fieldsite open, they also keep open the possibility to widen and rethink how they conceptualise and explain the very phenomenon they seek to study. By attending to how various research participants refer to the internet as a space during interviews and casual conversations during fieldwork, I was able to put forward an explanation of how DC became so compelling as a therapeutic practice while parting ways with existing insufficient and conflating explanations that centre around quality of information, mis/disinformation, and anti-science attitudes. Secondly, this case study shows how ethnographic studies on digital diaspora are necessarily multisited endeavours; even from a main base site – which in this case is a DC clinic in Ho Chi Minh City – ethnographic researchers always have to carefully listen to references about other places in order to chart a larger picture, that which gives us a fuller sense of the space that encompasses these interactions on the field. Finally, the case study demonstrates a larger need for researchers of digital diaspora to account for the agency of the technological network that affords the diasporic experience in all their material, sensorial, and contextual qualities. This should be factored into their research design by foregrounding the co-constitutive ways in which technologies produce and maintain spaces of flow – as well as enabling emergent imaginations of space that bear direct consequences on how diasporic actors live their lives.

Ethnographic design

This case study draws from my fieldwork over ten months between April 2019 and March 2020 in Vietnam and the US, where I observed and interviewed followers and practitioners of DC at their clinics, their homes, over videotelephone applications, and through their activities on Facebook. This study was reviewed and approved by the Human Ethics Sub-Committees (HESC) at the University of Melbourne. Among the interview participants was BQC himself, who has given written permission to not remain anonymous. Apart from BQC, all participants have been given pseudonyms in this paper. Six interview participants were recruited from a BQC-affiliated Facebook group, four were recruited from two other DC groups without affiliation with BQC, and eight more interview participants were recruited through chain-referral sampling (Bernard, 2017).

In July 2019, I began observing different DC Facebook groups for three months and keeping a research diary to map out the interaction dynamics of different members across these groups. Afterwards, I contacted participants and conducted in-person interviews at DC clinics, informed by the preliminary online chats with participants. During these interviews, I asked questions about the participants' encounters with DC, their decision-making process for practising DC professionally and as part of their healthcare regime, and their use of technology in relation to practising DC. The interview transcripts were analysed along with field notes from visits to DC clinics. While not all visits included participant interviews and not all participant interviews occurred in clinic settings, the data collected from clinic observation, online observation, and in-depth interviews were used to inform each other in reiterative cycles to refine the methodological techniques used and my evolving engagement with DC communities (Christ, 2010; Alvesson & Sköldberg, 2017).

Throughout 18 semi-structured interviews, all participants consistently referred to the internet as a 'place' as they described their involvement with DC. During one-on-one interviews and face-to-face visits to DC clinics, I delved into these expressions of 'space' and 'place', both directly and indirectly, as highlighted in interview transcripts and compared with field notes for a broader context. Four specific vignettes were chosen for analysis in the section below, each representing distinct spatial expressions that are most representative of those encountered in the collected data. The internet is a different kind of 'place' in each articulation, with some of the articulated places being more similar than others. These contemporaneous spatializations of the internet give rise to a non-biomedical milieu that enables anomalies to adapt and re-establish themselves as normal – sometimes even superior to biomedicine – in these emerging techno-social environments.

Diasporic spaces as material, sensorial, and contextual

1 The internet as an alternative techno-social site

Thu is in her early 40s; when I met Thu at her apartment-cum-clinic in a brand-new apartment complex inside one of Ho Chi Minh City's latest urban development zones, the door was kept open. As I entered and introduced myself, Thu casually nodded her head to acknowledge my presence and asked me to wait in the clinic room. In the middle of the clinic was a long wooden table, surrounded by plastic chairs on all four sides. There was also a lone hospital bed pushed against the top right corner of the room, under a large window with a view facing the Sai Gon River. As I sat down at the table on which dozens of tripods and smartphones were placed, I noticed that the clinic walls were filled to the brim with DC point chart posters that resemble acupuncture maps of pressure points on the body. I didn't mind the wait as there was a lot to take in; where point chart posters didn't fit, portraits of BQC and photos of him and his followers would enter to fill up what could otherwise have been some white space on these walls.

When Thu finally came in to begin our interview, I noticed that she had changed into her uniform – a light purple mandarin collared shirt with toggle buttons, which resembles traditional medicine doctor attire, and a large DC logo on her left chest. Two of her assistants quickly picked up the phones and tripods on the table to install a three-camera setup. I explained to her that it was her right as a participant to enjoy anonymity and to withdraw from the study at any point she so wishes. 'It's OK, I don't want to be anonymous. I livestream everything I do here at this clinic every day. Someone will watch it, that's how I stay connected with my clients [...] This humble clinic you see here is nothing compared to the splendour of our following on the net (*trên mạng*), but I guess you already know that,' said Thu.

And so our conversation began. Every time I came back to visit the clinic and talk to Thu's students and patients, our conversation would be live-streamed on Facebook.

These livestreams are automatically available as recordings for later viewing. I later learned that this almost obsessive documentation and broadcasting of everyday events has become a ritual of sorts for all members of her 'crew' – as well as other DC groups that I have been able to trace through Facebook and through my visits to BQC's own clinic. Thu explains to me that her compulsive livestreaming of DC activities is a form of record-making; it is as much a tactical undertaking as it is an act carried out precisely because Facebook has made the technology readily available:

> Why do I livestream my practice? I don't know, why not? I want to help people so the more people can watch what I do here the better. I help my patients get better every day, people should see it with their own eyes. As I practice the method on my patients, I also explain what I'm doing to the audience. They can watch and learn from what I do. This is like a bonus from my usual tutorial videos. Even if they cannot follow me when I treat my patients on video, they can look up the tutorial videos separately, using the keywords I mentioned in my demonstration livestreams. You know, you can re-watch all livestreams after the fact, and even search for them. A lot of people do. Sometimes it's a little difficult to find old livestreams, so we reshare it on our feed, or make new livestreams with slightly different storytelling. It's all really invaluable resource. The wisdom of Master Chau lives on in living videos like these.

This elaboration of livestreaming as record-making is attuned to the particular affordances of Facebook as a platform. Unlike other transient technologies, Facebook Live videos remain accessible indefinitely and can be viewed at any time, much like pre-recorded videos. Affordances refer to the ways in which technologies permit and limit potential behaviours, which can differ based on factors such as perception, skill, and cultural and institutional legitimacy (Davis, 2020). They can be best understood as relational properties that mediate action possibilities. While technologies cannot force people into doing things, it is through the mechanisms and conditions of their affordances that actions are encouraged or discouraged (Bucher & Helmond, 2017; Ettlinger, 2018, Hanckel et al., 2019).

Raman et al. (2018) analysed 3TB of Facebook Live data for patterns of global activity and found that most of the engagement with Facebook Live videos comes *after* the live broadcast – with engagement increasing about fourfold the day after broadcast. While the act of going 'live' in this context might be spontaneous, the digital traces it leaves behind are perceived not as a platform by-product, but as premeditated footprints that can be, and will be, drawn upon as a record. Record-making in this context is also space-grabbing. A record, once made, is 'stored' somewhere. While the metaphor of cloud computing – the imaginary space that our data immaterially reside 'in the clouds' – has become ubiquitous in English-speaking contexts, the same does not apply in this context. It seems imperative that a techno-social site be constructed to augment the limited physical space that DC takes up: one that remains visible and accessible insofar as the act of record-making never ceases. Record-making is not automatic record-keeping, however. Some level of archival orientation is necessary for record-keeping to occur – in other words, for records to be systematically managed and distributed. While the need for record-keeping (maintaining and indexing resources) is frequently mentioned, its practice is hampered by the impulse to 'go live' (you can always make more videos), perceived audience expectation (people want a constant stream of new videos or they would lose interest), and platform affordances that discourage effective subsequent knowledge retrieval and organisation (proprietary algorithms that prioritise livestream videos over other asynchronous broadcasts, achronological ordering of Facebook newsfeed). This trifecta conditions a dynamic sense of space-grabbing of the internet as constituted through interactions – of space as a product of interrelations constantly under construction. While record-keeping and by extension, systematic knowledge management requires mastery and ordering of space as an enclosure, excessive record-making is active open-ended space-making. This active space-grabbing through the abduction of time – of creating spectacular events to compete for audience attention – is driven by a techno-social precarity that, while having an enabling effect, is just as daunting as existing precarities facing DC.

> Yes, I sometimes worry about the authorities. They have made it difficult for us in the past. News media are also very biased against us, although recently they have softened their prejudice. The establishment sees us as unorthodox, as anti-establishment even. We're simply too avant-garde, even when Master Chau invented this method way back in 1980. But we are more ubiquitous than you might think. There are a lot of traditional medicine clinics who quietly practice DC. They don't advertise it publicly of course, they could lose their licenses for that. Master Chau has long given up on the need to be recognised by the Agency of Traditional Medicine Administration. They are simply too old-fashioned. We do our own thing now, without the need for approval from anyone. The internet is wonderful like that. It leads us to the true path for DC: anyone can be free to practice DC, teach DC, interpret and appropriate DC for their own needs, and all the while ignore the noise from the naysayers.

As such, the internet is seen as a venue in which the multiple and simultaneous trajectories of DC would coexist in their full liveliness, regardless of underlying coherence. While the spaces for mainstream biomedical and non-biomedical practices alike are made and representational – qualities that cannot be detached from their history and context – the technosocial space for DC is made of heterogeneous impromptu acts of broadcast that are at once generative and elusive.

2 Being alone with technology as being-in-space

Lam, who joined Thu's group from his hometown less than 200km south of Ho Chi Minh City, was 19 years old when I met him. I sat down with him at a coffee shop near the clinic to talk about his introduction to DC on a separate occasion. As soon as the conversation began, Lam was comfortable enough to share that in his final year of high school, he discovered DC because he had haemorrhoids at the time and 'Western medicine' was of no help to him. Out of frustration, he looked up haemorrhoid cures on the internet and followed a few DC tutorials on YouTube, which according to him was able to help him manage the condition. He later discovered the growing number of DC communities on Facebook, Thu's included, and found out that he could take a DC class directly with BQC if he would make the trip to Ho Chi Minh City and pay the VND5 million (roughly $220) tuition fee. When he finished high school, his score was not good enough for him to get into university; he figured that pursuing DC professionally – something he seemed to have a particular talent for – would set him apart from his high school friends, who were too busy 'chasing grades and studying what they hated'. With his parents' blessing, he moved to Ho Chi Minh City to embark on his journey as a DC professional.

> I couldn't talk to anyone about, you know, my haemorrhoids. I mean I eventually told my parents so they could take me to see a doctor, who was of little help … Since I was a kid I've always hated taking medications. It does no good to your body. I managed my condition by looking up cures on the internet, specifically using the keywords 'cures without medication' (*không dùng thuốc*). I found tons of such videos on Youtube, demonstrating how to cure my haemorrhoids by practicing DC techniques. One video, two videos, then ten, twenty … I can't recall exactly how many videos I've watched and practiced along to. Must be a lot. There were days when I completely lost track of time; I would start practicing to these videos after lunch only to stop for dinner. Then I would practice late into the evening before bed. I was hooked. You know what it's like to lose yourself to the internet, right? You get lost in there. Video led to video, then from YouTube I jumped to Facebook and Zalo (a popular Vietnamese messaging application). Just me and my phone, like this. Most of my friends waste their time on online video games, which is dumb. I learnt to cure myself and others from (*từ*) the net. It's a much more useful skill.

For Lam, the internet as a space might be characterised as providing the condition for therapeutic activities that generate time: time that can either be lost to wasteful online spaces or well spent being in spaces that help him with both his health condition and with finding a sense of purpose. The more time is spent on the internet, the larger the internet is envisioned as a space: to traverse and master a vast space, one needs to invest time. He maintained that his DC skills cured him of his haemorrhoids – the skills he later practised on his parents and extended family to help them with all their health problems to convince them of his unconventional career choice – and that the internet gave him an alternative space to explore where his abilities might best be invested. While Lam's individual experience of illness might have seemed isolated and isolating – that of an embarrassed young man practising along to obscure YouTube videos alone in his room – there is a great deal of sociability in Lam's experience of health subjectivity. This sociability is attributable to the spatiality with which the internet is thought of: Lam hardly noticed how time was passing as he immersed in a space of multiplicity – of the mobile technology he held in his palm, of the platforms he savvily navigated, of the connections he made through these platforms which eventually materialised into his mobility into the city. Lam's subjective experience with his health conditions was, therefore, not an internalised succession of sensations where the different points on his face were correctly massaged to somehow produce the effect of curing his haemorrhoids – an experience of pure temporality – but that of a relational nature and is therefore also spatial.

3 The internet as a conduit of healing power

Shortly after my first visit to Thu's clinic, she offered to introduce me to BQC at his residence. We sat down in his office on the second floor of his villa, tucked away in a small alleyway in Ho Chi Minh City. On the ground floor, people were sitting in the foyer waiting to see Cuong, one of BQC's sons, who is the practitioner on call for the day. The furniture would be moved around at night, Thu explained to me, so BQC could teach advanced DC classes in the same room where Cuong was diligently massaging his patients using strange-looking tools that can be bought in an adjacent room. BQC is remarkably lively for a man in his late 70s; he would, however, get quite grumpy when I would occasionally try to steer him back from his incessant reminiscence of the past. BQC has many wild stories to tell; what his stories lack in clarity and veracity, they make up for in intricate details, punctuated with random mentions of French philosophers and famous Vietnamese politicians. As we talked about his overseas following – which is largest in Italy and Spain – and how technology helped him stay connected with them, he assured me that digital technologies do more than just help him popularise his invention:

> I recently cured a whole family in Germany of back pain. Father, mother, and their son. You heard me, they're in Germany. Yes they are Vietnamese-Germans, the Vietkieus (overseas Vietnamese), but I cured them while they are in Germany, not while they are visiting here. You ask me how? It's through this phone right here. They were practicing my tutorials for back pain for weeks without any results, so they called me up using Zalo to ask for help. I asked the mum to show me the son's back on video. I then cured him through the videocall using my psychic energy through my palm. Just like I would if he was here physically. The phone carries my energy, you know. It travels through the internet, to Germany. You know, 4G technology these days, it's quite marvellous. He got better instantly. Only I can perform such a miraculous feat.

BQC initially attempted to legitimise DC through scientisation but later turned to the supernatural and spirituality to elevate his status to quasi-sainthood. This shift in approach coincided with his core group's adoption of new technologies. Although he does not manage his own social media presence, BQC uses his smartphone to make direct contact with his overseas following and record DC events. BQC and his followers view the internet as a space that can convey therapeutic energy and facilitate the therapeutic experience. In this view, the internet becomes a domesticated space: a space invented by science, but shaped in ways that accommodate the goals and desires of people engaging with that space. The internet is understood to possess special qualities that can be activated through certain rituals in particular events, which are both utilitarian and influenced by deeply held beliefs about well-being and technology. When the internet is successfully mobilised as such, it becomes more obscure and is viewed as a conduit for healing energies, reinforcing associations with the spiritual world. This view of the internet as a non-physical space that is embedded in webs of signification lends itself to spiritual interpretations and highlights the presence of the immaterial world that has been neglected by the biomedical gaze on the body.

4 The heterotopic internet and transnational health mobility

For Quang, a Vietnamese-American in his mid-60s who had been living in California since 1973, going online means engaging in a buffer space that allows him to draw on resources in the 'homeland' without having to compromise his ideological and political convictions. Having served in the South Vietnam army before Communist North Vietnam won the Vietnam War in 1975, Quang had vouched to never return to a Vietnam he refused to acknowledge as legitimate. I met Quang and his family of four, together with two of his best friends in the US, at Thu's clinic on one hot December afternoon in 2019. He had decided to spend his winter break in Ho Chi Minh City studying DC with BQC and frequenting Thu's clinic to both 'hang out' and get Thu to help with his wife's chronic fatigue. Quang had been diagnosed with prostate cancer four months before this trip; a family relative in the US recommended that he look up DC on Facebook and follow the livestreaming tutorials there to help with his illness. 'These people are legit, you know,' Quang told me firmly, 'I'm not one to be fooled. I know all this seems a little unconventional, but DC really works. I don't just uproot my family on a whim. My wife and I own a *phở* shop in Garden Grove, Orange County; it's not easy for us to leave the shop to our staff and come here for my treatment.'

Quang and I had a long conversation, which was of course livestreamed on Facebook, as we watched Thu perform various DC massage procedures on his wife, Xuan, who is in her early 50s. In another room, An, one of Thu's assistants, was performing other procedures on Quang's friends, Toan and Lan, who are in their 70s and 40s respectively. 'I had my doubts initially, as you can imagine. I live in the US; this kind of stuff is usually seen as quackery over there. I took the time to google 'dien chan' and 'scam' together. There was no result! But there were a lot of results for 'dien chan', in all kinds of languages, even Russian. All of these results sing praises about master BQC and Thu's group. That's how you know they are the real deal.' It was not my place to point out the flaws in his information appraisal strategy. Quang quickly became preoccupied with DC, practising it every day and closely following Thu's livestreaming videos on Facebook in a fashion similar to how Lam found himself 'lost' in a space of his own making. When he felt that the progression of his conditions required an intervention more radical than practising Thu's paid online tutorials, he decided to make the trip to Vietnam for the first time in 46 years.

> I'm not exaggerating when I say it's DC, Master Chau, and Ms Thu here that got me to go back to Vietnam. I no longer have any family here … I might as well be a stranger in a strange land. The community I found in DC reminds me of a Vietnameseness that is rooted in traditional wisdom. Traditions, as you know, are free of politics. When you innovate a tradition, such as a medical tradition, you bring Vietnameseness to the world. Traditions have to be innovated where they started, so to speak. And then brought overseas. You cannot have real Vietnamese traditions in the US, for example. We're Vietnamese over there, but we also had to be Americans. It's a different kind of community. Once I'm in Master Chau's advanced DC class, I found a community almost immediately. People practiced the techniques on me and let me practice on them. It's easy when the culture is already there and you just become part of it. My treatment with DC is even more effective when I'm here.

Conclusion

This chapter has made the case for digital diasporas as empirical research sites that are not naturally occurring, ready-made fields of events – but rather as complex arrangements of difference that require, and allow for, the development of new vocabulary beyond that of 'culture' and 'society' as autonomous entities. As sites of ongoing enactments of sociality, digital diasporas are made up of constantly shifting movements and connections that assemble and disassemble on the basis of shared identity, common meaning, and bounded sovereignty that is nevertheless always subject to transcension. They are deeply embedded in the complex terrains of social lives, and so the main challenge for researchers is not to carve out a diasporic space that has a clearly defined beginning and end, but rather to carve out the relief figures of diasporic connections against the background that gives them depth. As the digital diaspora assemblage evolves, some connections will recede to the background while others will become more prominent, and new connections are made. Researchers of digital diaspora are interested not in the study of wholes-of-culture; they are after the dynamic and open-ended nature of diasporic networks. To fully commit to this approach, researchers need to articulate their theory of space in their ethnographic practices; this chapter encourages researchers to consciously factor space into their empirical data collection by not only taking note of how research participants bring up space in conversations, but also proactively render their spatial conception of the fieldsite open, so as to keep open the possibility of re-thinking the phenomenon they seek to study. Researchers of digital diaspora should also factor in the agency of technologies that mediate and mediatise diasporic experiences in all their material, sensorial, and contextual qualities – which would help enrich and texturise their ethnographic accounts. In the next chapter, we will discuss digital ethnography as an emergent approach that has been developed to examine how humans feel, create, respond, and engage with digital technologies – and what this approach can do for digital diaspora studies.

References

Alvesson, M., & Sköldberg, K. (2017). *Reflexive methodology: New vistas for qualitative research*. SAGE.

Ardèvol, E., & Lanzeni, D. (2017). Ethnography and the ongoing in digital design. In L Hjorth, H Horst, A Galloway, & G Bell (Eds.), *The Routledge companion to digital ethnography* (pp. 474–483). Routledge.

Axel, B. K. (2004). The context of diaspora. *Cultural Anthropology*, 19(1), 26–60.

Aziz, A. (2022). Rohingya diaspora online: Mapping the spaces of visibility, resistance and transnational identity on social media. *New Media & Society*, 14614448221132241.

Beaulieu, A. (2017). Vectors for fieldwork: Computational thinking and new modes of ethnography. In L Hjorth, H Horst, A Galloway, & G Bell (Eds.), *The Routledge companion to digital ethnography* (pp. 55–65). Routledge.

Bernard, H. R. (2017). *Research methods in anthropology: Qualitative and quantitative approaches*. Rowman & Littlefield.

Brubaker, R. (2005). The 'diaspora' diaspora. *Ethnic and Racial Studies*, 28(1), 1–19.

Brühne, J., & Kuhlmann, H. (2022). 'Extended diaspora': On communitization phenomena in the digital age. *Journal of Global Diaspora & Media*, 3(Textures of Diaspora and (Post-) Digitality: A Cultural Studies Approach), 19–37.

Bucher, T., & Helmond, A. (2017). The affordances of social media platforms. In J Burgess, A Marwock, & T Poell (Eds.), *The SAGE handbook of social media* (pp. 233–253).

Burrell, J. (2017). The fieldsite as a network: a strategy for locating ethnographic research. In L Hjorth, H Horst, A Galloway, & G Bell (Eds.), *The Routledge companion to digital ethnography* (pp. 77–86). Routledge.

Candidatu, L., & Ponzanesi, S. (2022). Digital diasporas: Staying with the trouble. *Communication, Culture and Critique*, 15(2), 261–268.

Candidatu, L., Leurs, K., & Ponzanesi, S. (2019). Digital diasporas: Beyond the buzzword: toward a relational understanding of mobility and connectivity. In J Retis, & R Tsagarousianou (Eds.), *The handbook of diasporas, media, and culture* (pp. 31–47). John Wiley & Sons, Inc.

Champion, E. M. (2021). *Rethinking virtual places*. Indiana University Press.

Christ, T. W. (2010). Teaching mixed methods and action research: pedagogical, practical, and evaluative considerations. In A Tashakkori, & C Teddlie (Eds.), *SAGE handbook of mixed methods in social and behavioral research* (vol. 2, pp. 643–676).

Davis, J. L. (2020). *How artifacts afford: the power and politics of everyday things*. MIT Press.

Ettlinger, N. (2018). Algorithmic affordances for productive resistance. *Big Data & Society*, 5(1). 10.1177/2053951718771399.

Graham, M. (2013). Geography/internet: ethereal alternate dimensions of cyberspace or grounded augmented realities? *The Geographical Journal*, 179(2), 177–182. 10.1111/geoj.12009.

Hanckel, B., Vivienne, S., Byron, P., Robards, B., & Churchill, B. (2019). 'That's not necessarily for them': LGBTIQ+ young people, social media platform affordances and identity curation. *Culture & Society*, 41(8), 1261–1278. 10.1177/0163443719846612.

Mitra, A. (2008). Working in cybernetic space: Diasporic Indian call center workers in the outsourced world. *South Asian technospaces*, 205–224.

Nedelcu, M. (2019). Digital diasporas. In R Cohen, & C Fischer (Eds.), *Routledge handbook of diaspora studies* (pp. 241–250). Routledge.

Nguyen, D., Arnold, M., & Chenhall, R. (2021). The internet as non-biomedical milieu: Production of alternative health techno-social spaces and the persistence of marginalised medical practices. *Health & Place*, 70, 102583. 10.1016/j.healthplace.2021.102583.

Ponzanesi, S. (2020). Digital diasporas: Postcoloniality, media and affect. *Interventions*, 22(8), 977–993. 10.1080/1369801X.2020.1718537.

Raman, A., Tyson, G., & Sastry, N. (2018). Facebook (A) live? Are live social broadcasts really broad casts? In *Proceedings of the 2018 world wide web conference* (pp. 1491–1500).

Sai Gon Giai Phong (1988, November 17). Cuba: Phổ biến phương pháp diện chẩn của Việt Nam trong mạng lưới y tế gia đình [translation of title]. Sai Gon Giai Phong. http://baochi.nlv.gov.vn/baochi/cgi-bin/baochi?a=d&d=UyRi19881117&e=------197-vi-20--341--img-txIN-c%E1%BA%A3i+t%E1%BA%A1o+----1975-.

Sai Gon Giai Phong (2012a, April 10). The hazy truth behind the dien chan method, part 1: telling the real from the fake. *Sai Gon Giai Phong Online*. https://www.sggp.org.vn/mo-ho-lieu-phap-dien-chan-bai-1-thuc-hu-dien-chan-post130343.html.

Sai Gon Giai Phong (2012b, April 10). The Hazy Truth behind the Dien Chan Method, Part 2: Unproven Efficacy. *Sai Gon Giai Phong Online*. https://www.sggp.org.vn/mo-ho-lieu-phap-dien-chan-bai-2-hieu-qua-chua-duoc-minh-dinh-post130407.html.

Sai Gon Giai Phong (2012c, April 13). A follow up on the Dien Chan method: Legal actions will be taken on unauthorised medical treatments. *Sai Gon Giai Phong Online*. https://www.sggp.org.vn/phan-hoi-tu-bai-viet-mo-ho-lieu-phap-dien-chan-s e-xu-ly-viec-kham-chua-benh-trai-phep-130463.html.

VNExpress Online (2003, August 29). The Dien Chan method lacks scientific foundations. *VNExpress Online*. (Reprinted from Lao Dong Newspaper). https://vnexpress.net/phep-chua-benh-bang-dien-chan-chua-du-co-so-khoa-hoc-2255609.html.

Vu, V. H. (2020). *An invention by a world-renowned doctor*. Dien Chan Viet Online. http://www.dienchanviet.com/dien-chan/vu-van-hoi/bui-quoc-chau-nha-phat-minh-dien-chan.

Wimmer, A., & Glick Schiller, N. (2002). Methodological nationalism and beyond: nation–state building, migration and the social sciences. *Global Networks*, 2(4), 301–334. 10.1111/1471-0374.00043.

Witteborn, S. (2019). Digital diaspora. In R Tsagarousianou, & J Retis (Eds.), *The Handbook of diasporas, media & culture* (pp. 179–192). Wiley Blackwell.

5

DIGITAL ETHNOGRAPHY FOR THE DIASPORA

Introduction

Digital ethnography refers to a range of approaches to conducting ethnographic research on digital culture and practices which foregrounds the digital as not only enabling and mediating, but also constitutive of, social lives. Early digital ethnographic works by trained anthropologists were invested in understanding social lives in emergent virtual worlds and virtual communities. Tom Boellstorff's *Coming of Age in Second Life: An Anthropologist Explores the Virtually Human* (2008), for example, explored the thriving alternate universe of *Second Life* where residents create communities, buy property and build homes, go to concerts, meet in bars, attend weddings and religious services, buy and sell virtual goods and services, find friendship, and fall in love – by conducting more than two years of fieldwork living among and observing residents of *Second Life* as he would any non-virtual culture and social group. Assuming the avatar of 'Tom Bukowski' and applying the methods of anthropology throughout his time in *Second Life*, Boellstorff (2008) was able to rigorously theorise about many facets of human life, including issues of gender, race, sex, money, conflict, and antisocial behaviour, the construction of place and time, and the interplay of self and group. Doing ethnography in an online context, as such, was a methodologically innovative avenue for anthropologists to rethink classical topics in their discipline; Boellstorff (2008) ultimately argued that in some ways humans have always been virtual and that virtual worlds in all their rich complexity build upon a human capacity for culture that is as old as humanity itself. Elsewhere, Bonnie Nardi's *My Life as a Night Elf Priest: An Anthropological Account of World of Warcraft* (2010) studied play as an active aesthetic experience by conducting fieldwork in China and the US over the course of three years. By conducting fieldwork in Beijing cafes and other parts of China to study the practice of 'modding' – where players create and distribute software modifications that extend the game as an act of creative collaboration – Nardi (2010) offered an analysis of the relationship between digital code, creativity, aesthetic experience, and human activity. These early digital ethnographic studies sought to establish virtual worlds as valid venues for cultural

DOI: 10.4324/9781003336556-8

practice in order to make claims on how they resemble and differ from other forms of culture, and to document and theorise on how virtual worlds themselves are vital places of social interaction and cultural activity.

Online communities were also of central concern to scholars working with digital ethnography in the late 1990s/early 2000s. Nancy Baym's book *Tune in, Log on: Soaps, Fandom, and Online Community* (2000) is a classic ethnographic study of online communities in the tradition of communication studies. Baym (2000) showed how verbal and non-verbal communicative practices create collaborative interpretations and criticism, group humour, interpersonal relationships, group norms, and individual identity by extensively studying a female-dominated soap opera fan community. As digital ethnography becomes increasingly practised – and embedded – in cognate disciplines such as media studies, internet studies, digital sociology, and science and technology studies (STS), the subject of digital ethnography also proliferates beyond that of the virtual world. Veronica Barassi's *Activism on the Web: Everyday Struggles against Digital Capitalism* (2015) studied the everyday tensions that political activists face as they come to terms with the increasingly commercialised nature of web technologies by conducting ethnographic fieldwork in the UK, Italy, and Spain. By examining the tensions created by communication processes and networked individualism, corporate surveillance and data mining, fast capitalism and the temporality of immediacy, Barassi (2015) explored the complex dialectics between digital discourses and digital practices, the technical and the social, and the political economy of the web and its lived critique. danah boyd's *It's Complicated: The Social Lives of Networked Teens* (2014) studied how teenagers communicate through social media services such as Facebook, Twitter, and Instagram by conducting a decade of original fieldwork interviewing teenagers across the United States. By exploring tropes about identity, privacy, safety, danger, and bullying, boyd (2014) painted a nuanced picture of what it means to grow up networked and argued that paternalism and protectionism could seriously hinder teenagers' ability to become informed, thoughtful, and engaged citizens through their online interactions.

Researchers of diaspora have also engaged with digital ethnography to understand how diaspora formation – as well as the enactments of diasporic lives – are changing at the intersection of technology and society. Weaving politics, media, and diaspora together to theorise the nation as network, Victoria Bernal's book *Nation as Network: Diaspora, Cyberspace, and Citizenship* (2014) mobilised fieldwork conducted on Dehai.org, Awate.com, and Asmarino.com as the core of the Eritrean transnational public sphere over a decade to develop the concept of 'infopolitics'. Infopolitics advances theories of sovereignty and understandings of the internet by foregrounding the management of information as a central aspect of politics, while revealing the ways that sovereignty and citizenship are being reconfigured and reproduced by means of the internet. Situating online activities on these websites not as a feature of diaspora, but as part of the configuration of Eritrean nationhood, Bernal (2014) argued that the Eritrean digital diaspora is a public and communal space that is a staging ground for ideas and practices that have no offline counterpart, either inside or outside Eritrea. Based on this analysis, Bernal (2014) developed useful concepts such as diasporic citizenship to capture the ambiguous political membership in the nation of the people in the diaspora, and sacrificial citizenship to describe the political culture of the Eritrean state and its insistence on its citizens to die for the nation.

Elsewhere, Ulla D. Berg (2015) mobilised fieldwork conducted from 1998–2005 in rural communities of the Mantaro Valley in Peru's central highlands, and following migrants from that area, in the Peruvian cities of Huancayo, Lima, and the U.S. destinations of Miami, Washington, D.C., and Paterson, New Jersey to trace the experiences, practices, and imaginaries of transnational migration and mobility among Andean Peruvians. Her book *Mobile Selves Race, Migration, and Belonging in Peru and the U.S.* (2015) explored the ways in which migration is mediated between the Peruvian Andes and the United States by focusing on the forms of sociality and belonging that technological mediations enable, while consciously contributing to debates in anthropology about affect, subjectivity, and sociality. By analysing how communicative practices that use visual, rhetorical, and material resources and cues are central to the way Andean Peruvians over the course of migration become increasingly conscious of their own social positioning within a larger social and racial order and within the multiple social relationships they maintain and create across borders, Berg (2015) showed that the larger constraints of the migration process constantly prompt migrants to communicate to others – elite Peruvians, people in migrants' home towns and urban neighbourhoods in Peru, U.S. immigration officials, employers, and wider publics – an image of what one is and who one wishes or is expected to be.

The sample of works above should give readers a sense of the growing importance of digital ethnography in social research, as well as the diverse empirical domains that benefit from engaging with digital ethnography. This chapter discusses digital ethnography as an approach to studying digital diaspora. The chapter starts with a discussion on the notion of the multisituated researcher as an extension of the discussion on multisited ethnographic research in Chapter 4 – and as a point of reflection for researchers of digital diaspora, who themselves might be diasporic actors with multiple accountabilities to multiple communities of practice, such that disciplinary reproduction is not the only stake for many emergent practitioners in the field. The chapter follows with a discussion on recent developments in digital ethnography, with a focus on how ethnography as a method has evolved to accommodate the growing need to rigorously study digital life. Because the digital has become inseparable from the everyday, digital ethnography as method and as a research perspective is not only attuned to the study of virtual worlds, but also the material and social worlds we inhabit alongside technologies. The chapter discusses how digital ethnography as an emergent approach that has been developed to examine how humans feel, create, respond, and engage with digital technologies – and what this approach can do for digital diaspora studies. The chapter concludes with a case study on the temporal structures of diasporic lives by drawing on fieldwork conducted in Vietnam and the US on the livestreaming practices of *diện chẩn*, an emergent unregulated therapeutic method among the Vietnamese diasporas.

The multisituated researcher and the politics of diaspora

Writing about the complex diasporic itineraries and investments that ethnographers often carry with them as they conduct fieldwork – and what they might mean for ethnographic praxis – Kaushik Sundar Rajan (2021) developed the concept of 'multisituated ethnography' in direct conversation with, and as an extension of, multisited and multilocale ethnography (see Chapter 4). Acknowledging that ethnography exceeds the discipline of anthropology (and is thus widely practiced across history, sociology, political science, literary studies, creative arts, legal studies, media studies, STS, and indeed beyond academia itself), Kaushik Sundar Rajan nevertheless situated ethnographic practices against not only the time and place of the ethnographic situation, but also the inheritance of ethnography as a scholarly practice rooted in disciplinary and institutional norms that are grounded in colonial, masculinist histories. Contending with this inheritance, he argues, means that ethnographers should be attuned to an ethnographic ethos whose stakes are not about making the strange familiar or the familiar strange – as the anthropological mantra often goes – but about staying 'with the trouble of the complex epistemic and institutional relationships between reproduction and kinship, kinship and alterity, and diaspora and reproduction that lie at the heart of ethnographic practice and disciplinary pedagogy' (Rajan, 2021, p. 187).

In this sense, the notion of a multisituated ethnography is meant to supplement that of multisited or multilocale ethnography; it is meant to encourage researchers to cultivate a stance, or disposition, that allows for building institutional arrangements that foster a fuller expression and inhabitation of the multiple investments and accountabilities that ethnographers have towards the communities of practice with whom they inhabit their practice. Increasingly, ethnographers are themselves diasporic actors who often travel away from the communities they study to acquire professional membership in university communities of practice – the professionalised metropolitan academe – before shuttling back and forth between these two locales for fieldwork, while opening themselves up to the possibility of extending their work to other locales and commitments as they diligently follow their actors and connections in the field. The diasporic ethnographer, therefore, 'invariably has to code-switch between the disjunctures of "familiarity" and "strangeness" that often exist between her own personal, intellectual, and political biographies and that of a metropolitan disciplinary history that is professionalising her' (Rajan, 2021, p. 4). The Other in ethnography is no longer someone to be ethically accountable to, but also the source through which to make any claims to knowledge about the general, the structural, and the systemic. This is at the heart of what Rajan (2021) means by his articulation of a multisituated ethnography – as feminist and de-colonising in its ethos, and as committed to the creative and evocative potentials of humanistic and artistic ways of knowing and doing.

But the diasporic question in multisituated ethnography is not just about international students and their complex identities and addressees. For Kaushik Sundar Rajan, the politics of diaspora in ethnographic praxis speaks to all manners of intersectional intellectual, political, and biographical trajectories – and commands the politics of a disciplinary disavowal of its colonial inheritance through an actual, demographic trend towards decolonisation by its diasporic practitioners. In practicing multisituated ethnography, researchers are committed to cultivating conditions for the proliferation of possibilities and of ethnographic modalities – even when these possibilities and modalities are partial, provisional, and frictioned with respect to one another. Rajan (2021) articulated the practice of multisituated ethnography through four main dimensions: scale, comparison, encounter, and dialogue – against, around, and through which researchers work to contend with the difficulties of the colonial and masculine inheritances of ethnography. Through this articulation, Kaushik Sundar Rajan showed how the epistemic consequences of what it means to be a diasporic researcher in the metropolitan university are tied not only with the transnational politics of diaspora, but also with the problem-spaces, disciplinary paradox, and praxis of ethnography.

Rather than developing multisituated ethnography as a programmatic response to the colonial and masculinist inheritance as well as the realities of changing epistemologies in ethnography, Rajan (2021) fleshed out the idea of multisituated ethnography as an ideal of personal and autobiographical sensibilities and modalities of ethnographic practice – a situated reflection of what ethnography has come to mean for him as a diasporic ethnographer coming from India, trained in the US, having conducted fieldwork in both places, and at the time of the publication of his book developing an affinity with South Africa. In the question of scale, Rajan (2021) reflected on the disjuncture between the theoretical aspirations of anthropology that demand conceptualisation across site and scale, and the actual norms and forms of ethnographic fieldwork and narration that are valorised in anthropological research projects. The objects of anthropological theory tend to be abstract and operate across sites and scales, such as 'global capital' or 'human rights', or 'governmentality' and 'financialisation'; it is in this sense that the literal multiplication of ethnographic sites and cross-national comparison only goes some way towards bridging the theoretical ambitions and the actual practice of ethnography. Ethnography as an approach privileges a mode of producing knowledge which uses the descriptive, the particular, and the exemplary in order to produce generalised or systemic understandings; the stakes around articulating a multisituated ethnography that empowers researchers to stay with the complexities of the ethnographic situation becomes contingent upon the very situated ways in which researchers activate their own stakes in the matter. For Kaushik Sundar Rajan, designing for a multisituated ethnography that meets the challenge of scale firstly involves a rigorous elucidation of context – historically, locationally, and situationally – as *ethnographic* in and of itself, rather than as background to the work of ethnography. As such, the work of ethnography is not just about tracing, but also *making* connections as articulation of structural relations across locales. The objects, practices, or phenomena under study by means of ethnography – be it infrastructure, health, diaspora, or technology – do not reveal themselves through ethnographic practices per se, but rather become ethnographic questions of the questions about themselves. This is a function of the interactivity and reflexivity between the researcher and the world they inhabit: by thinking about multisituated ethnography not as a literalist methodology (as in, putting oneself in as many ethnographic situations as possible for the sake of scale), but rather as a conceptual topology by considering the relationship between the stuff of the world we wish to empirically describe and the stuff of the world we wish conceptualise. In conceptualising digital diaspora, for example, researchers might empirically describe how diasporic actors use their mobile phones to read news about home, browse social media, livestream events, play online games, edit photo collages, send remittances through mobile apps, have family discussions on group chats, wait for a consultation with doctors at home to go live, and so on. These activities do not happen against the background of a ready-made digital diaspora; they are the stuff with which digital diasporas are made of. Scaling out the particular and contingent encounters that one has as an ethnographer into systemic and structural claims is a matter of breaking the binaries of the particular and universal, experience-near and experience-far, -emic and -etic, invisible and visible, material and abstract, micro and macro, there to here. While offering no single formula or mechanistic technique on how to do this, Rajan (2021) insists that for an ethnography to be meaningful, it has to straddle these tensions, simultaneously inhabiting and deconstructing them. The ethnographer tacks back and forth between actors' and analysts' categories, in ways that have to be constantly accountable to the former but that cannot be reduced to them. Between what can be observed by the ethnographer (such as practices, rituals, events) and the abstraction of systems (which are not tangible things in themselves, yet nonetheless have tangible effects that can be observed) lies the multisituated disposition and sensibilities of care, of attending to their informants' agencies, of conceptualising the local site as always deeply embedded in national, regional, and global relations of power and production, and of being attuned to a present as thoroughly historicised.

In the question of comparison, Rajan (2021) urged researchers to critically assess ethnographic norms and forms that go beyond the monologic dimensions of sociological analysis and theory making as critical objectification of Western epistemologies. This translates to an ongoing articulation of the study of political modernity in and from the postcolony – one that simultaneously reifies and decentres conventionally colonial metropolitan-periphery relations. This can be done both in the literal sense of doing comparative ethnography, and thinking comparatively about sometimes similar, sometimes very different, sometimes deeply related cultures. In practice, ethnographic design often follows the ethnographer's analytic agenda, even as it is responsive and responsible to the empirics of the fields it studies; for researchers of digital diaspora, carrying out the analytical work of comparison means looking beyond any seemingly universalistic technological affordances that structure diasporic interactions and situating these observable interactions firmly against the historically and geographically contingent formations of diasporic networks. This brings us to the question of the ethnographic encounter, which Kaushik Sundar Rajan discussed in relation to photographic praxis. The ethnographic encounter, he argued, can never be innocent with all its potential and actual intimacy with the Other; at the heart of this encounter is the politics of sight that structures the relationship between stranger and kin, whose morality cannot be reduced to the conceptual transformation of binaries such as 'evocation' as somehow more virtuous than 'representation', or 'transference' more virtuous than 'objectification'. Even with institutional codes of ethics in place, which are meant to get researchers to tabulate the risks and benefits of fieldwork to safeguard against any gross violations, the ethnographic encounter can never be purely contractual. Here, the idea of multiple accountabilities towards multiple communities of practice comes back to remind us that the politics of ethnography lies beyond the simple construction of the triangulated relationship between ethnographer, native informant, and subsequent reader. Practicing multisituated ethnography means staying attuned to the reproductions and circulations of these relationships; it is not about the politics of voice, or giving voice to the Other, but rather about creating institutional space and disciplinary affordance for diasporic political commitments – including that which will queer the project of disciplinary reproduction itself. In the case study at the end of this chapter, I show how by attending to the temporal structures of the ethnographic encounter in my fieldwork, I reached across the language of anthropology, media studies, and STS to construct an ethnographic account of diasporic enactments through and with livestreaming technologies. This is but one example of how one might refunction the norms and forms of ethnographic practice in order to create new kinds of conversational spaces for its enactment.

In the last dimension – dialogue – Rajan (2021) considered multisituatedness as an explicitly dialogic practice, one that questions not only who the ethnographer is conversing with, but also where she is conversing from: the experiential milieu out of which her ethnographic attentiveness has been sculpted. Encouraging ethnographers to engage in a diverse range of ethnographic modalities, Kaushik Sundar Rajan conceptualised the ethnographic problem-space as emerging alongside a host of investigative, interpretive, and creative modalities that complement, supplement, and interrupt the knowledge that ethnography produces. Reflexivity in multisituated praxis is a thick, situated reflexivity – not a casual choice, or self-indulgence – so that the ethnographer's itineraries can become visibly part of the *mise-en-scène* of the complex, interrelated, and hybrid worlds she seeks to elucidate. Situating the ethnographic dialogue against what Holmes and Marcus call para-ethnography in their seminal essay *Cultures of Expertise and the Management of Globalisation: Toward the Re-functioning of Ethnography* (2005), Rajan (2021) urges ethnographers to reconsider the native informant as interlocutor and collaborator – and through such an analytical move not only reconfigures the object-subject relationship, but also rescripts the informant's place within the ethnographic encounter. Even when the ethnographer is observing and not actively conversing – as much of ethnographic work necessarily is – Rajan (2021) insisted that the ethnographer should also make active and relational intervention in the scene of the ethnographic encounter in order to render it dialogic. Discussing how to do this effectively, Rajan (2021) nominated the para-site as the most scripted para-ethnographic encounter, where conversation between key interlocutors and the ethnographer is designed to follow certain debates staged in a fashion that is unfamiliar and generative to both the ethnographer and all interlocutors involved. In this configuration, the ethnographer does not determine the trajectory of the conversations – as traditional interviewing techniques would dictate – but rather determine their *mise-en-scène* of the ethnographic situation. The para-site as a space of stage encounter – an ethnographic third space – can help unsettle notions of romantic authorship and disciplinary production that often go unquestioned. As an experimental ethnographic modality, however, Rajan (2021) admitted that while the para-site is but one mode of situated dialogic engagement that could be strategically viable at a certain stage of a project, within certain kinds of interactional milieux – it cannot be the totality of an ethnographic project. This is particularly potent when applied to the theoretical and empirical concerns of researchers of digital diaspora. In what ways are digital diasporas 'third spaces' – and in what ways are they field sites? What is to be gained – and lost – when we make such analytical moves? In what ways can digital diaspora researchers render their ethnographic encounters dialogic, following the multisituated agenda – and produce the kind of ethnography that not only challenge, but also transcend disciplinary reproduction and institutional norms?

In the next section, we will consider another relatively recent development in ethnography – digital ethnography – in particular relation to the study of digital diasporas. If multisituated ethnography as articulated by Kaushik Sundar Rajan is an attempt at capturing and articulating something akin to an epistemological shift in ethnographic praxis at the intersection of anthropology and STS – that which foregrounds the multiple investments and accountabilities that diasporic ethnographers have towards the communities of practice with whom they inhabit their practice – digital ethnography is an emergent, open-ended, multidisciplinary project that deals with the central question of how to practice ethnography so that researchers can fully acknowledge and account for the ways in which digital media and technologies are part of the everyday and more spectacular worlds that people inhabit. By championing flexibility and adaptability through adopting the pragmatic view that digital ethnography should continuously bend itself to become the best suited tool for the situation and the most resonant framework for understanding, digital ethnography in some ways free itself altogether from the concerns around institutional norms and disciplinary reproduction that Rajan (2021) invested considerable effort into subverting, queering, and extending. Emerging at the intersection of anthropology, sociology, media studies, STS, human-computer interaction, informatics, information studies, human geography, and cultural studies, digital ethnography is often practiced in explicitly multidisciplinary ways, in conjunction with other social science and humanities methods, employing digital tools for data collection and analysis, and embracing creative and evocative forms of conducting and communicating its knowledge.

Digital ethnography and the experience and practice of digital diaspora

Reflecting on the depths of what ethnography must confront given the ways in which digital technologies have come to construct the shape of social lives, Mario L. Small (2022) remarked that even though telephones – a communication technology that precedes the internet – changed as many aspects of everyday life as the internet and smartphone have, they nevertheless did not change the basic expectation about what an ethnographer was supposed to do. 'Ethnography survived the telephone unscathed' (p. 478), he quipped, because the phone call was private, intentional, and exclusive, such that the ethnographer could at best observe one party, and not the interaction between both – and as such did not constitute an ethnographic space in the same way that the internet does. It is no longer possible to escape digital ethnography, argued Small (2022), because the proliferation of digital institutions across all facets of social life has rendered the ecological conditions of the physical location less central to the work of the ethnographer, all while ensuring that digital and non-digital action are inextricably linked. Taking a step further to make the case for digital ethnography, Small (2022, p. 480) insisted that 'the classical fieldworker will have no greater claim to authenticity than the coffeeshop ethnographer,' especially after the way that the COVID-19 lockdowns have changed people's expectations around what it means to be social. Not just as a means to help researchers better study the social phenomena of concern, digital ethnography also allows us the space for new conceptual categories that might force us to rethink the social world – and as a result, alter our understanding of the world as a whole.

In a seminal text on digital ethnography, Pink et al. (2015) outlined five key principles for doing digital ethnography: multiplicity, non-digital-centricness, openness, reflexivity, and unorthodox. The multiplicity principle recognises in explicit terms that there is more than one way to engage with the digital and encourages researchers to be guided by not just theoretical frameworks specific to their disciplines, but also the needs and interests of their research partners, stakeholders, and participants. What the digital constitutes might indeed differ as the researcher enters 'the field'. As I will show in the case study below, it became clear to me during the time I spent with the *diện chẩn* group that what necessitates the digital in their everyday practice was not a coherent or comprehensive digital strategy to attract and engage with a global audience, but rather a more fundamental structure in which the digital rescues and recuperates downtime for practitioners and followers of this method. Researchers of digital diaspora, therefore, should render their working understanding of the digital open and flexible as they do the notion of the field. More than just a matter of reflexively questioning the salience and performances of the digital in field interactions, or augmenting digital engagements with research participants through non-digital, keeping in mind that the shape of the digital looks differently to different people allows researchers to get closer to the underlying issue they seek to study. This also relates closely to the second principle, namely non-digital-centricness. This principle encourages researchers to think of media and technology as always already embedded in a wider set of environments and relations – and as such inseparable from other activities, technologies, materialities, and feelings through which they are used, experienced, and operate. In practical terms, this principle encourages researchers to not ask research participants directly about their media or technology use – as in the crude formulation of 'what role is technology playing in your life?' – but rather engage in the everyday routines and activities that colour their participants' social and digital worlds. Often this would involve working with concepts other than that of the digital, but that which allows the digital to come through in better focus (e.g., downtime, in the case study below).

The openness principle encourages researchers to think of digital ethnography as a fundamentally collaborative process. In this sense, digital ethnography is open to influences not only by identified collaborators (including research participants and interlocutors), but also non-academic, non-traditional knowledge practices such as speculative design or arts practice. This openness is not enacted in the simple form of adding as many diverse actors as possible into the ethnographic design mix, but rather in thinking critically and reflexively about how these diverse ethnographic modalities can come together to co-produce knowledge in genuinely insightful and interesting ways. How might engaging with arts-based techniques help research participants better articulate their lived experience with automated technologies in ways that are not only insightful to the researcher, but also useful to the participants themselves? How can ethnographic knowledge be communicated back to communities of practice in ways that help them better make sense of the challenges and joys of living with technologies? Can research participants be encouraged to engage with ongoing practices of ethnographic knowledge production by drawing on creative methods in their everyday lives?

This ties to the fifth and last principle of digital ethnography as being committed to being unorthodox, to acknowledging and seeking out ways of knowing what are otherwise invisible and unanticipated by the more formally constituted approaches. Digital ethnography attends to the embodied and sensory experiences of engaging with new technologies, machines, and devices – a domain of inquiry routinely explored by creative and artistic practices – by explicitly drawing on, collaborating with, and learning from, these practices. Sarah Pink and John Postill (2017), for example, explored transient student migration by conducting video ethnography of Indonesian students in Australia in their everyday practice of doing laundry. Conceptualising the status of being an international student as 'student migrant' in order to refer to the ways that moving country has implications for everyday life that are both specific to being a student, but that are also concerned with the process of migration from one locality to another,

Pink and Postill (2017) posited that researchers interested in the experiences of migration should try to immerse themselves in the private worlds where migrants encounter and improvise with new domestic technologies, products and processes. Arguing that doing the laundry is a mundane yet essential part of how temporary student migrants encounter the materiality of home, finance, technology, infrastructure, weather, and their social worlds in a new place, Pink and Postill (2017) followed eight Indonesian students in Melbourne with the sensory video ethnography and in-depth interview approach developed where they interviewed these students in an unstructured fashion before conducting a video tour of the parts of their homes and digital media technologies that participants felt comfortable showing, and a re-enactment of their laundry processes – from seeking items to be washed through to hanging items out to dry. What they were interested in through this process is how it felt (both sensorially and emotionally) to do laundry in Melbourne, the ways participants engaged with, improvised with, and appropriated technologies for laundry, how these experiences were framed by a particular weather world, and how they connected with ways of knowing that were aligned with traditional Javanese culture and activity. This research design is particularly insightful and novel against the background of a large number of studies on student migrants/diasporic students that focused on the social networks, social media, religion, housing, and language skills of international student migrants – all of which are important topics that can benefit from the foregrounding of sensorial and emotional experiences of digital ethnography.

It is important to note that the realities of practicing digital ethnography are not without complications, quandaries, and discomforts. Riffing off Gary Alan Fine's (1993) essay *Ten lies of ethnography*, Gabriele de Seta (2020) outlined three lies of digital ethnography through the archetypical figures of the networked field-weaver, the eager participant-lurker, and the expert fabricator. Contending that researchers have to perform a gentle choreography of professionalism and persuasion every time they are prompted to speak about their research and fieldwork – a practice which requires researchers to carefully massage their research description according to the needs of the moment – de Seta (2020) interrogated the discursive strategies, performative masks, and illusory identities that researchers likely confront in their thinking, speaking, and writing on digital ethnography. Following Gary Alan Fine, de Seta (2020) embraced the 'uncomfortable truths of the trade' and acknowledged how certain unavoidable instances of lying are integral to what it means to practice digital ethnography. This does not mean that digital ethnography is some form of fakery, or that digital ethnographers routinely cheat in their practice; rather, what this 'lying' accomplishes is allowing for the choices made by ethnographers to accommodate to varying working conditions, textual forms of output, and requirements of secrecy part of the ethnographer's accountability to their research participants.

The lie of the networked field-weaver, explains de Seta (2020), comes in effect as the digital ethnographer relies on the 'field as networks' metaphor to render the trajectory of their fieldwork open-ended only to immediately cut away at the straws they're constantly grasping to stay on track and to retain a legible focus – rather than experiencing the expansive movement of branching out as the metaphor would otherwise suggest. Digital ethnographers deliberately ignore leads, decline offers to further socialisation, prune the outbranching connections they encounter, switch directions, wrap up their study as funding runs out, seal off information outside of a project's stated scope in the interest of timely completion. The field is also often weaved more than once during the life course of a project, which could linger long after the digital ethnographer has completed fieldwork. As you will note in the case study presented below, I have mobilised more or less the same set of field interactions that I included in the case study in Chapter 4 to make a completely different argument on the temporal structure of downtime, which gives shape to the livestreaming practice of *diện chẩn*. What I have done here is pulling back some of the same data into new configurations – and by so doing modifying my working sense of 'the field' to accommodate the discursive positioning of my research, based on both the rhetorical necessities of my intended audience as well as the specific aspect of diasporic life I want to explain. As such, the networked field-weaver not only cuts away at the branches as they emerge, but perhaps more importantly also pulls together otherwise disparate, fragmented threads that at first glance do not mesh together. The networked field-weaver's lie, as such, is essential to producing good, useful, and legible digital ethnography; this lie becomes problematic, however, when the digital ethnographer takes for real the notion that she might be the only person who could oversee the multi-sites she chooses, and only she alone can see the patterns she assembles. Being explicit and honest about how one weaves their field-network beyond a formalised exercise in self-reflexivity would help readers – and all communities to which the digital ethnographer is accountable – capture what would otherwise be 'distorted into clarity' (Law, 2004, p. 2).

The lie of the eager participant-lurker is precisely that she is not one. Digital ethnographic accounts are often written up to give the impression that the ethnographer is 'an eager participant-lurker: a master of all modes of participation, portrayed as impossibly co-located across multiple fieldsites, surveying digital media use from a vantage of carefully crafted presence' (de Seta, 2020, p. 88). Reflecting on his research on WeChat, de Seta (2020) remarked that as an already savvy user of Chinese platforms from his extended time living in China, not much changed when he started fieldwork for his project. In his own words:

> I was still browsing websites, scrolling through social media feeds, chatting with friends, liking their posts, commenting on news stories, watching and listening to content shared by my contacts, collecting samples of interactions and writing fieldnotes to wrap up daily observations and encounters. The only thing that changed was that I wasn't sitting at my Hong Kong office desk, but rather wandering in Shenzhen, Wuhan, Shanghai, or Beijing, meeting friends I had not seen for a while, spending time with my partner, playing at experimental music concerts, and sitting in cafes with interviewees. (p. 85)

Being candid about the realities of doing digital ethnography brings a certain kind of texture to the traditional genre of the ethnographic account, which often encourages researchers to flatten her own persona into the apologetic figure of an eager participant-lurker. If we take seriously the tenet that the digital is central to social lives, then it follows that as researchers we also engage in a wide range of modes of participation that are tightly connected to social dynamics and technological affordances. These modes of participation do not simply recede to the background to accommodate for the privileged ethnographic participation; in exploring the creation of intersubjectivity as a fluid out-come of an ongoing ethnographic engagement that does not cease when the research project finishes, researchers can better situate their reflexivity against the *mise-en-scène* of everyday digital culture.

The lie of the expert fabricator, finally, has to do with the presentation of ethno-graphic accounts to ensure the integrity of the relational and situational research ethics negotiated on the field, as dependent on different digital media contexts. This is what Kaushik Sundar Rajan alluded to when he invoked the para-site in his discussion on dialogue and ethnographic practice to subvert and transcend the necessary authorial decisions made in the writing up of an ethnographic account. Even when the ethno-graphic author grounds her writing on extensive datasets, copious amounts of field-notes, and collections of traces, ethnographic accounts produced by digital ethnographers inevitably draw on a narrow selection of inscriptions, often thoroughly edited, translated, scrambled, rephrased, anonymised, cropped, and selectively blurred and collated according to a constellation of ethical, argumentative, and aesthetic authorial decisions (de Seta, 2020). This ethnographic fabrication is inextricably linked to the idea of expertise; as the digital ethnographer embraces her role as editor, translator, and fabricator of multimedia and multimodal vignettes, of composites of events, identities, and inscriptions (de Seta, 2020), she implicitly asserts her own competence and knowledgeability over the sociotechnical context she presents to her audience. The essential lie in this arrangement is how it hides how most ethnographic research is actually grounded on a patchy process of discovery – a messy interaction between the researcher's inquiries and the patient explanations of research partici-pants. de Seta (2020) calls on digital ethnographers to build transparency into their ethnographic accounts by not hiding behind expert and ethical fabrications, and instead fully acknowledging the roles of the ethnographic interlocutor in the co-production of ethnographic knowledge.

In the case study below, I present an account of how downtime structures the digital practice and experiences of *diện chẩn*. There are three aspects of doing digital ethno-graphy that I wish to reflect on with this case study. Firstly, I want to reflect on how I came to think about downtime as the concept through which to bring forth the digital in *diện chẩn* livestreaming practices. I had already finished the bulk of my fieldwork in the US when I became stranded in New Haven, Connecticut at the beginning of the COVID-19 pandemic in March 2020 as borders started to shut down. My PhD fellowship was scheduled to end in May, and yet I had no way of returning to either Vietnam (my country of citizenship) or Australia (where I did my doctoral studies). As I spent my days social distancing from everyone else in a graduate house in the increasingly deserted college town of New Haven, I found myself living through a kind of directionless time-in-waiting that can only become transformed through me obsessively clinging onto my digital devices. Going through my fieldwork notes on my tablet, I began to connect with what my *diện chẩn* interlocutors confided in me on a much more personal level. By reflecting on how the kind of downtime that can be experienced anywhere and by anyone can be recuperated with digital technologies, I started thinking about lives-treaming not simply as a technologically enabled practice, but as already embedded in wider temporal structures that it seeks to capture, disrupt, and rescue. By reflecting on the everyday routines and activities that colour our collective digital social worlds, I became better attuned to why and how *diện chẩn* became such a compelling therapeutic practice through and with livestreaming technologies.

Secondly, I want to reflect on how being included in many of the livestreaming videos made by the *diện chẩn* group during my fieldwork made me rethink my pre-fieldwork assumptions about what it means to 'go live'. The easy availability and accessibility of livestreaming technologies and platforms that enable livestreaming mean that one does not need a special occasion to go live. Once an activity reserved for sporting events, concerts, speeches, or momentous ceremonies – all of which require meticulous planning and coordination – livestreaming now happens whenever someone with a mobile phone and internet connection decides to go live. While I was busy going through the checklist instituted by my ethics committee to ensure I was doing all I could to protect my participants' anonymity, all our interviews were conveniently livestreamed on Facebook as part of how livestreaming has become a fabric of the digital sociality at the *diện chẩn* clinic. Finally, I want to reflect on the importance of taking cue from research participants by listening carefully to what they have to say, so that preconceived concepts and notions that guide a researcher into the field never takes precedent over what actually goes on in the field. As I tried to get a better sense of why *diện chẩn* practitioners and followers would invest so much time and energy into producing and consuming livestreamed videos, I found that all my research participants were letting me in on the answers precisely through what they were not telling me. Coming into the field, I expected to collect testimonies on how *diện chẩn* practitioners have shifted to organise their practice around the affordances of livestreaming. As I soon found out, livestreaming as an activity is secondary to whatever ends up happening in a day's work for these practitioners; the cameras switch on when patients visit the clinic for treatment regardless of whether the patients give their consent or not, and on slow days these practitioners would livestream *diện chẩn* tutorials to pass the time. The relationship between time, technology, and body becomes that of an extremely intimate nature, as I will show in the case study below.

Case study: theorising the temporal structures of diasporic lives

This section reports on the methods and key findings in Nguyen (2021) as a case study for doing digital ethnography by attending carefully in the paradoxes of the digital. The section introduces the domain-specific literature that the paper contributes to by briefly introducing its theoretical framework and its research questions before reporting in detail on the practical methodological choices made in its analysis. The case study demonstrates how researchers of digital diaspora can theorise about the temporal structures of digital diasporic practices in new and interdisciplinary ways by productively engaging with the paradox of liveness as articulated in technical literatures. The case study highlights the tensions that exist in the ways technical literatures and media studies conceptualise certain properties of the digital, and encourages researchers to mobilise digital ethnography towards clarifying, explaining, reconciling, and forging a way forward for these tensions. In thinking about the temporal structures of diasporic lives through digital ethnography, researchers not only contribute to the much-needed empirically informed theorisation of the digital, but also become well-positioned to develop new, interdisciplinary theoretical frameworks for the study of digital diaspora. By juxtaposing time-as-algorithmic against time-as-lived through the livestreaming practices of *diện chẩn*, an emergent unregulated therapeutic method (also see Chapter 4), the case study shows how different enactments of liveness on Facebook Live recalibrate downtime so that the body can reconfigure its being-in-time. The temporal reverberation of downtime and liveness creates an alternative temporal space wherein social practices that are shunned by the temporal structures of institution and society can retune and continue to thrive at the margin of these structures and at the central of the everyday.

Background to case study

In a computationally extensive study, Raman et al. (2018) analysed 3TB of Facebook Live data for patterns of global activity only to question whether the platform is 'truly' live, or indeed can even be considered a 'broadcast' service at all. They question whether Facebook Live can be considered a truly 'live' or 'broadcast' service based on their finding that most engagement with Facebook Live videos occurs after the live broadcast. During the live broadcast, videos in their dataset received relatively low engagement, but engagement counts increased significantly one day after the broadcast. Specifically, on average during the live broadcast, videos in their dataset receive 6.7 likes, 8.4 comments, and 0.54 shares; one day after broadcast, the engagement counts jump to 29.84 likes, 16.33 comments, and 1.33 shares (Raman et al., 2018). Lamenting that because as much as 41.5% of all Facebook Live videos were never watched, the researchers suggest locally storing the video content on the broadcasters' mobile devices until viewers arrive to save network bandwidth and battery consumption. This recommendation, if taken up by Facebook Live, triggers a fundamentally different model of content circulation on the platform: one that resembles an on-demand service, where livestreaming videos without an audience are set free from the mobile devices that house them only if these livestreams fail to satisfy the conditions that make them 'live' in the first place.

What to make of this apparent paradox? From a technical point of view, liveness seems to be synonymous with simultaneity: are there people watching at the other end, as the video is being recorded? Yet for liveness to come about, instantaneity also needs to be at play: a video trapped inside a mobile device simply cannot 'go live.' At the heart of what motivates the researchers of this study lies the unexamined intimate relationship between 'real-time' and sociality – be it to the event/performance, or to people – that colours the experience of liveness. The meaning of 'live' is always contrasted to and informed by the 'non-live'; it is because of this that the conditions under which liveness comes into being deserve disentangling. Probing into the apparent paradox of liveness on Facebook Live could tell us something important about our relationship with time and the technology that mediatises that relationship. An ethnographic inquiry into the lived experience of liveness shows that liveness oscillates between instantaneity and simultaneity as it takes on multiple iterations contingent on platform affordances. Liveness as instantaneity transforms downtime as moments of disjuncture, and simultaneity provides temporal structure for the tactical persistence of practices that are out of sync with the rhythms of institutions and society.

In the sections that follow, I report on the results of my fieldwork in 2019 on DC practice in Vietnam and the US, where I observed and interviewed followers and practitioners of the method at their clinics, their homes, and over videotelephone applications. I also interviewed BQC himself, although his interview is not reported here. In so doing, I explore how downtime is drawn upon by interviewees to explain their engagement with the method as well as the technologies that make the method readily and instantly accessible to them. I have given my interviewees pseudonyms to protect their identities. As DC moved away from the legal and scientific battles for recognition, it increasingly moves towards the internet, where its tactile propositions find resonance with the temporality of the mobile technologies on which it travels, as it intervenes in the self-maintenance of the body and interrupts the temporalities of everyday life.

The main vignettes from my fieldwork are Thu, Lam, and Quang. Each of these vignettes illustrates distinct experiences of multiple liveness, medial self-time, and technology. Thu commits to a career in livestreaming DC therapy tutorials to transform downtime as a larger episode of personal crisis; yet the increasing intensity and scope of what she understands as appropriate for livestreaming is a transformation of downtime as a disjuncture of the moment. Lam consumes these therapy livestreams to transform the downtime that comes with experiencing illness; his subsequent decision to pursue livestreaming as a producer both resolves the need to overcome downtime as residing between the action and inaction of everyday life and as an opportunity to translate a digital practice into urban mobility. Quang's experience with consuming therapy livestreams is similar to Lam's, yet his transnational mobility is tied to his need to recuperate time for a sick body that has been made to wait, away from home.

The durée of downtime

I met Thu, who is in her early 40s, at her apartment-clinic located in a new apartment complex within Ho Chi Minh City's urban development zones. When I arrived, she left the door open and casually nodded her head to acknowledge my presence, then asked me to wait in the clinic room. The clinic had a long wooden table in the middle of the room, surrounded by plastic chairs on all sides. There was also a single hospital bed pushed against the top right corner of the room, under a large window with a view facing the Sai Gon River. The clinic walls were covered in DC point chart posters that looked like acupuncture maps of pressure points on the body. As I sat down at the table on which dozens of tripods and smartphones were placed, I noticed that the clinic walls were filled to the brim with DC point chart posters that resemble acupuncture maps of pressure points on the body. I didn't mind the wait as there was a lot to take in; where point chart posters didn't fit, portraits of BQC and photos of him and his followers would enter to fill up what could otherwise have been some white space on these walls.

After waiting for a bit, Thu finally came in for our interview. I couldn't help but notice that she had changed into her uniform - a light purple mandarin collared shirt with toggle buttons that looked like traditional medicine doctor attire, with a large DC logo on her left chest. Two of her assistants quickly moved to pick up the phones and tripods on the table to set up a three-camera arrangement. I made sure to explain to her that she had the right to remain anonymous and withdraw from the study at any time. 'It's OK, I don't want to be anonymous. I livestream everything I do here at this clinic every day. Someone will watch it, that's how I stay connected with my clients,' said Thu.

Our conversation with Thu began, and each time I returned to the clinic to talk to her students and patients, we would livestream the conversation on Facebook. I later discovered that this almost compulsive documentation and broadcasting of everyday events has become a kind of tradition for all members of her crew. The crew livestreams everyday events on Facebook simply because they can, and the possibility of an audience is sufficient justification for broadcasting an event, regardless of its quality or purpose. Because Facebook automatically archives livestreamed recordings after the livestream has concluded, Thu and her crew consider 'going live' as a significant record-making activity. Thu shared that she holds a university degree and has worked for several international companies for nearly a decade, which allowed her to practice her English and gain exposure to foreign cultures. Her radical turn to DC, a '100% Vietnamese therapeutic method', coincided with major disruptions in her life: disillusionment with a 9 to 5 job that did not pay well, marital problems, the birth of her son, and a sudden but deeply felt need to reconnect with Vietnamese culture. She quit her job, moved back in with her parents, and waited for her life to turn around. As would be the case, Thu's mother happens to be one of BQC's original followers since the 1980s. Her mother, Ms Tuyet, would frequently travel south, leaving her and her father behind in Northern Vietnam during the 1980s and 1990s, to accompany BQC on various DC-related trips overseas.

> It was fate … when I felt the most stuck, unseen, and unfulfilled, my mother said maybe it was time I turned to DC. Not only can I take care of myself with this craft, I can even take care of thousands of other people all around the world. They watch me demonstrate, then they practice this [method] on themselves. They watch me practice it on other people, then they do the same on their families and friends. If it works, their families and friends would then practice it on themselves, then on their friends … it's an ever-expanding circle of care.

Thu's practice of DC has filled her downtime with activity on two different levels. During a significant episode of personal crisis – this is downtime on a lifetime scale – Thu turned to DC to pulsate the flatlining temporal rhythm in which she found herself.

> This is a very dignified craft … before I became involved in DC, I would never dream of commanding a police officer or telling him to follow my instructions. A police officer! Can you believe it? One of my patients is a policeman. I told him to lie down on this bed and take off his shirt so I could perform DC on him to help with his cervical vertebrae. I livestreamed it on Facebook of course, you can always check it. People like him would never have given me the time of day in my previous life as a nobody. DC has completely changed my life for the better.

Thu needed a sense of purpose; a purpose that can be achieved by doing something instantly, so as not to feel left behind. There is a certain indignity to being someone who waits in a culture of the instant. To sit around waiting is to be out of sync with modernity, with the habit of velocity that dictates how one should live their life; it is to be out of sync with time itself.

Downtime also comes with the experience of illness. When I met Lam, he was only 19 years old and had joined Thu's group from his hometown located less than 200km south of Ho Chi Minh City. During our conversation at a nearby coffee shop, Lam shared with me his introduction to DC. He said that he discovered DC in his final year of high school when he was suffering from haemorrhoids, and Western medicine was of no help to him. He became frustrated and looked up haemorrhoid cures on the internet, which led him to a few DC tutorials on YouTube. He found that these tutorials were able to help him manage his condition. Later, he discovered the growing number of DC communities on Facebook, including Thu's group. He learned that he could take a DC class directly with BQC if he made the trip to Ho Chi Minh City and paid VND5 million (roughly $220) as tuition fee. Lam's score was not good enough for him to get into university, so he decided to pursue DC professionally. With his parents' support, he moved to Ho Chi Minh City to start his journey as a DC professional, believing that this path would set him apart from his high school friends who were 'chasing grades and studying what they hated.'

> My main job in Ms. Thu's group is packing and sending DC tools to customers, as well as designing visual materials for social media … I won't lie, I was disappointed at first. I expected to be practicing DC on patients right away, curing and helping people. After all, I took my classes with master BQC himself. But now I know that working these tasks gives me more confidence, more experience in communicating with people. Ms. Thu pays me well for someone with no diplomas to speak of. The people we help, especially people we met in a pagoda in Bình Định province on our DC mission trip, appreciate and respect us. I could not ask for a better job right out of high school.

If downtime for Lam is tied to both the experience of an illness that damaged his self-confidence and the anxieties of getting out of school without a clear direction, downtime for Quang, a Vietnamese-American in his mid-60s who had been living in California since 1973, is a matter of mortality. Having served in the South Vietnam army before Communist North Vietnam won the war in 1975, Quang had vouched to never return to Vietnam on ideological grounds. I met Quang and his family of four, together with two of his best friends in the US, at Thu's clinic on one hot December afternoon in 2019. He had decided to spend his winter break in Ho Chi Minh City studying DC with BQC and frequenting Thu's clinic to both 'hang out' and get Thu to help with his wife's chronic fatigue. Quang had been diagnosed with prostate cancer four months before this trip; a family relative in the US recommended that he look up DC on Facebook and follow the livestreaming tutorials there to help with his illness. 'These people are legit, you know,' Quang told me firmly, 'I'm not one to be fooled. I know all this seems a little unconventional, but DC really works. I don't just uproot my family on a whim. My wife and I own a *phở* shop in Garden Grove, Orange County; it's not easy for us to leave the shop to our staff and come here for my treatment.'

Quang and I had a long conversation, which was of course livestreamed on Facebook, as we watched Thu perform various DC massage procedures on his wife, Xuan, who is in her early 50s. In another room, An, one of Thu's assistants, was performing other procedures on Quang's friends, Toan and Lan, who are in their 70s and 40s respectively. 'I had my doubts initially, as you can imagine. I live in the US, this kind of stuff is usually seen as quackery over there. I took the time to google 'dien chan' and 'scam' together. There was no result! But there were a lot of results for 'dien chan', in all kinds of languages, even Russian. All of these results sing praises about master BQC and Thu's group. That's how you know they are the real deal.' It was not my place to point out the flaws in his information appraisal strategy. Quang quickly became preoccupied with DC, practising it every day and closely following Thu's livestreaming videos on Facebook.

> The greatest thing about this method is that it allows you to be in charge of your own body, your own health … Western medicine takes away all that power from you. You go to the hospital and they let you die there. No dignity whatsoever. Obama made us pay premium prices for nothing. Trump is no better. Here you can practice on yourself, on your family, on your friends. Wherever it hurts, you control it with your own hands. There is true power in that ability.

The body cannot wait: the oscillation between instantaneity and simultaneity in liveness

If there is indignity in waiting around, there is enchantment in waiting for an object of desire: on the other side of waiting stands the promise of meaningful connections and the abolition of boredom. If waiting around is *eigenzeit* being out of sync with the times of others, waiting as enchantment ensures that recalibration is always an option. Liveness as simultaneity is the promise of this enchantment; as instantaneity, the promise of dignity: after all, to show respect is to not keep someone waiting. This double recalibration of liveness could act as an efficient time transformer as it enables social arrangements to veer towards the integration and assimilation of different temporal horizons and speeds. Making a sick body wait can be a painful and humiliating ordeal. A body recuperating itself from waiting by extending its primary tactual exploration part – the haptic hands – to touch, scroll, pinch, slide, press, massage, wave, and gesticulate, is a body quite literally taking matters into its own hands. We cannot help sensing tactual sensibilities through our skin any more than we can help the passing of time; or as Bergson (2002, p. 216) would put it, 'It is we who are passing when we say time passes.'

> Thu is livestreaming all the time, so it's rare that I would run out of materials to learn. If not a tutorial then she'll be streaming her performance on someone, or introducing new tools that I can buy, or talking about DC in general. There is a significant time difference between Sai Gon and California, so sometimes I watch her livestreams after the fact. Doesn't matter to me. She's very charming, Ms Thu. She has what it takes for this craft. She knows how to speak to people. Sometimes I watch the same video again and again; I would learn something new every time.'

Xuan smiled at me as she sat up straight on the designated treatment bed so An could perform a range of massage techniques that were supposed to help Xuan with her chronic fatigue. She was never as involved as her husband Quang in the practice of DC but remained supportive. 'Quang is the kind of man who becomes obsessed really quickly. Whatever he chooses to pursue, he invests all his time and attention.' 'Was that how he pursued you?' – giggled Thu, as she asked Xuan to move a little to the left.

> Sure he did. But seriously, he is engrossed in this thing. Always on his phone watching videos, at the *phở* shop, at home, picking up the kids, before going to bed … One time

I even asked if he's forgotten that he had a wife! He apologised by giving me these DC massages to help me with my back problem. It really did help. When he decided that we would come back to Vietnam so he could learn more about this method, I gave him my full support. We're all here, aren't we?

As downtime becomes recalibrated, it expands and transforms the structure of the everyday. Technology as an option with which to live through downtime has the capacity to weave downtime into the very fabric of time itself. As far as Thu and Lam are concerned, downtime has been permanently recalibrated; as long as they can livestream their works instantly, every day, they are producing digital materials that reiterate their newfound social prestige and rescue them from a previous kind of downtime: downtime as dwelling endlessly in a material world that turns against them, so that their bodies are not so much material facts, but manifestations of duration. 'It is we who are passing when we say time passes' (Bergson, 2002, p. 216). Recalibrated downtime is a kind of medial *eigenzeit* that relieves the body from feeling its own weight. Without the technologies that transform, repurpose, and objectify time, we become time's vessel: the sick body feels this most emphatically. For Quang, every moment he otherwise would have spent worrying and feeling helpless is now an opportunity to engage both in a bodied practice that is already instantly available, and in a transnational community built on the mobilization of downtime. Quang's recalibrated downtime is living through medial *eigenzeit* that yearns for reconnection with the body in the moment and on its own terms. Quang told me when Thu and An were out of the room:

Look, I'm not delusional. I know there is no absolute cure to my cancer. But if there is something out there that helps, even just a little, of course I'd try. And so far, these exercises have really worked. I feel much better, bit by bit, every day. You know, me in front of my phone, in my palm, like this. She [Thu] presses (acupressure) point number 12 three times – I press point number 12 three times … It's like I'm practicing it with her. There is a kind of genius in that simplicity.

Liveness is not reducible to instantaneity, however. A recording of a live event is different from a non-live video recording, not only because it is experienced as such, but also because it requires specific temporal coordination from the live-streamer and the audience, as well as particular technological affordances to facilitate that liveness. Thu explained that as her audience base of *Việt Kiều* – a colloquial term referring to overseas Vietnamese – expands, she becomes more strategic about what kind of content to stream at which point of the day.

Most of these *Việt Kiều* live in California. That means our afternoon is their morning. I've realised that most of them are up and online at around 4pm our time. I try to schedule as many appointments with my clients in the afternoon around that time as possible, so they can wake up and see me in action. If that's not possible, I always prepare to give a live tutorial around their common problems, like back pain, migraine, heart diseases. Sexual problems are also common. Or problems with fertility. I listen to their problems, and I show them a way to solve these problems. Their feedback is key to this process.

This feedback loop not only informs what should go live, but it also informs the temporality of liveness. Encountering live videos is increasingly becoming a dominant feature in the Facebook experience: in 2017, one out of five videos on Facebook are live videos (99firms, 2020). The liveness of Facebook Live videos owes itself in large part to its platform logic: even though Facebook itself is not seen as a 'live' platform, its News Feed is commonly associated with liveness, both in popular discourse and by the platform itself (van Es, 2017). A Facebook Live video appearing on a user's News Feed therefore enjoys this double sense of liveness; the instantaneity of the event is relative to when the event becomes visible to the user. This crucial timing depends not on universal time, but on Facebook's News Feed algorithm – a techno-social artefact built on implicit and concealed implementations of network temporality. Facebook algorithms are proprietary and 'black-boxed'; while it is common knowledge that these algorithms are continually tweaked, a complete dissection of these algorithms is neither available nor possible. Some *metatext* about Facebook algorithms is available, however. van Es (2017) reviewed the principles of temporal organising of content on Facebook through two main algorithms, EdgeRank (for page post and status update personalisation) and GraphRank (for application recommendation personalisation, less relevant in our discussion). EdgeRank calculates an affinity score between Facebook users based on the number of interactions between the user initiating the connection (edge) and the viewing user, the weight of each edge type, and time decay – a measure that takes into consideration how long ago an edge was created. The newer the edge, the more relevant it is to the users involved: the higher the likelihood of the associated content to appear on News Feed. This elaboration of network sociality has an implicit built-in temporality, one that favours both instantaneity and simultaneity, although perhaps not of equal measures – there is no transparent way to tell. Given the dynamic nature of Facebook networks, due both to the constant modifications made to platform algorithms and to the trajectories of human relationships, the organisation of Facebook content is likely to oscillate between these two temporal qualities. The reverberation of simultaneity and instantaneity on all levels of this constellation of liveness – from platform temporalities to the multiscale downtime that carves out a *space of participation* where techno-cultural, economic, and extra-legal forces converge, and the everyday reflections and tactics of uses from users (or *user responses*) – circumscribe the domain of liveness on Facebook.

Downtime, recalibrated: enactments of liveness and their effects

Previous sections have shown that liveness is variously enacted rather than existing in a pure form that can readily be contrasted against the 'non-live'. These enactments oscillate between simultaneity and instantaneity: the decision to 'go live' on Facebook is made because it can be done instantly, just as the decision to watch a live recording is made because it is instantly available. Thu would livestream everything happening in her clinic, simply because she had the necessary equipment at hand, and Quang would watch these livestreams multiple times as they were instantly available on his phone. Simultaneity is not very important when it comes to livestreaming on Facebook Live, as it is already built into the platform's features. When there is an attempt to coordinate simultaneity in Facebook livestreaming, it does not set live events apart from non-live events, but from other live events of different temporality. Thu's shift in livestreaming strategy, in response to her audience in California, adds a temporal layer to her practice of livestreaming, rather than restructures it to align with real-time. A live broadcast is significant because it is enacted as live, rather than being in line with universal time through technologies that represent time in a fixed manner. The experience of liveness does not occur in time as such, but in the enactment of itself, which is contingent on the various elements of liveness coming together. These different experiences of 'live' help the body reconfigure its being-in-time and fill the temporal voids with shifting webs of temporal sociality. These varying experiences of 'live' not only help the body reconfigure its being-in-time as instantly and readily replenished with meaning and activity, but also fill these otherwise temporal voids – experiences of pure temporality – with shifting webs of temporal sociality.

Even though user enactments of liveness might contrast with the kind of liveness enacted by engineers and computer scientists (see again, Raman et al., 2018), liveness is not fragmented into many. Engineering live streaming into Facebook – a platform whose temporalities already oscillate between simultaneity and instantaneity thanks to the algorithms that organise its content and condition its participation – is a task inevitably informed by this oscillation. There is a reverberation of this oscillation on three different levels: on the live streaming feature itself, within the platform, and in the downtimes of its users. The result of this reverberation is a radical recalibration of downtime with technology: one in which the body as sensory central reaches out to its mobile technology to actively reconfigure its being-in-time. In the specific case examined here, the body also took a leap of faith to direct its haptic and tactile explorations onto itself – living with technology in a recalibrated downtime has allowed the body to recognise itself also as an entity to be acted upon, to be reworked and renovated.

While not all recalibrations will entail this reworking of the body, the case examined here has outlined how this leap of faith is made possible not so much because of the newness or uniqueness of the practice *content* or the liveness *event*, but rather because of the *temporal reverberation* of downtime and liveness as enactments of the body and its technology. In other words, it is neither the uniqueness of DC as a method nor the novelty of livestreaming strangers massaging each other using odd-looking tools that allow the practice of DC to gain traction: the content and the event of liveness are expendable, owing themselves to the temporal reverberation in time, technology, and the body. This reverberation also carves out a temporal space for practices and enactments that are shunned by temporal structures of institutions and society to thrive: after all, downtime comes into being in the fractures between the mismatches of temporal regimes. Downtime and the insistence it puts on the body, which could mobilise both the production and consumption of livestreaming, are an often overlooked aspect of livestreaming practices. In the 'constellation of liveness' framework developed by van Es (2017), for example, liveness is a particular interaction between institutions, technologies, and people. It is through the analytical lens of downtime, as discussed in this article, that the mechanisms of how this interaction takes place can be traced. Enactments of liveness on Facebook is a result of temporal reverberations of downtime and liveness as enactments of the body and its technology, the result of which is a tactical temporal space that thrives at the margin of institutional temporal structures.

And yet there are inherent limits to this reverberation. The duration of liveness and downtime alike is limited, much like the finitude of a lifetime. Recalibrating downtime in service of rescuing the present moment through engaging with liveness is but one of the many ways in which the body can reflect on its embedding in its lifetime – a reminder that clock time marches forward, unwavering. Quang recognised that his prostate cancer would eventually outrun his exercises, Thu recognised that she could not out-generate tutorials for the growing conditions of her client base, and Lam recognised that at some point he would outgrow being the delivery boy for Thu's group. Any temporary unity between the self and its being-in-time at the behest of technology is intervened by its conflicted longing for the durability of human existence, which reminds the self of its ultimate precarity and eventual decay. Constant recalibration of downtime through enacting liveness is a tactical cheat that is able to seize onto little triumphs in between the cracks of temporal clashes, but unable to keep any of them. The alternative temporal space created through the reverberation of downtime and liveness continues to thrive at the margin of more enduring temporal structures, even as it becomes front and centre in the everyday lives of those relying on it for survival.

Conclusion

This chapter has introduced the tenets of digital ethnography as they relate to digital diaspora research and discussed how the adaptive nature of digital ethnography can accommodate the complexities of the diasporic situation as an elastic theoretical framework. Increasingly, researchers of digital diasporas are themselves diasporic actors whose commitments and accountabilities to various communities of practice are not only changing the expectations around ethnographic practices but also around the production of ethnographic knowledge. It is no longer the case that ethnographers can make a legitimate claim on producing knowledge through their research participants; it is more accurate to say that all ethnographic knowledges are produced with interlocutors in the field, whose expertise not only helps give shape to the digital ethnographer's inquiries, but also allows her to renovate discipline-specific legacy concepts that are often taken for granted. The key strength of digital ethnography lies in its capacity to enable researchers to approach the study of technologies in ways that speak to all manners of intersectional intellectual, political, and biographical trajectories – rather than black boxing technologies as technical systems devoid of social shaping – and in its commitment to a rigorous elucidation of the historical, locational, and situational context as *ethnographic* in and of itself, rather than as background to the work of ethnography. Digital ethnography's commitment to being unorthodox also gives digital diaspora researchers the mandate to engage fully with technologies in innovative and creative ways – not only by deploying digital technologies during fieldwork, as in the case of video ethnography, but also by mobilising, deconstructing, and theorising digital trace data in ways that allow us to fundamentally rethink technological processes. In the next chapter, we will turn our attention to the workings of automated systems as infrastructure for the emergence of digital diasporas and consider the methodological implications of these systems for researchers of digital diaspora.

References

99firms (2020). Facebook live statistics. Retrieved from https://99firms.com/blog/facebook-live-stats/#gref.
Barassi, V. (2015). *Activism on the web: Everyday struggles against digital capitalism*. Routledge.

Baym, N. K. (2000). *Tune in, log on: Soaps, fandom, and online community.* SAGE.

Berg, U. D. (2015). *Mobile selves: race, migration, and belonging in Peru and the US.* NYU Press.

Bergson, H. (2002). Duration and simultaneity. In K A Pearson, & J Mullarkey (Eds.), *Henri Bergson: key writings.* Continuum, London, UK.

Bernal, V. (2014). *Nation as network: diaspora, cyberspace, and citizenship.* University of Chicago Press.

Boellstorff, T. (2008). *Coming of age in second life: an anthropologist explores the virtually human.* Princeton University Press.

Boyd, D. (2014). *It's complicated: The social lives of networked teens.* Yale University Press.

De Seta, G. (2020). Three lies of digital ethnography. *Journal of Digital Social Research*, 2(1), 77–97. 10.33621/jdsr.v2i1.24.

Fine, G. A. (1993). Ten lies of ethnography: Moral dilemmas of field research. *Journal of Contemporary Ethnography*, 22(3), 267–294. 10.1177/089124193022003001.

Law, J. (2004). *After method: Mess in social science research.* Routledge.

Nardi, B. (2010). *My life as a night elf priest: An anthropological account of World of Warcraft.* University of Michigan Press.

Nguyen, D. (2021). Can't wait to feel better: Facebook Live and the recalibration of downtime in tending to the body. *Media, Culture & Society*, 43(6), 984–999. 10.1177/0163443721100345.

Pink, S., & Postill, J. (2017). Student migration and domestic improvisation: Transient migration through the experience of everyday laundry. *Transitions: Journal of Transient Migration*, 1(1), 13–28.

Pink, S., Horst, H., Postill, J., Hjorth, L., Lewis, T., & Tacchi, J. (2015). *Digital ethnography: Principles and practice.* SAGE.

Rajan, K. S. (2021). *Multisituated: Ethnography as diasporic praxis.* Duke University Press.

Raman, A., Tyson, G., & Sastry, N. (2018, April). Facebook (A) live? Are live social broadcasts really broad casts? In *Proceedings of the 2018 world wide web conference* (pp. 1491–1500).

Small, M. L. (2022). Ethnography upgraded. *Qualitative Sociology*, 45(3), 477–482. 10.1007/s11133-022-09519-1.

van Es, K. (2017). *The future of live.* Polity Press, Cambridge, UK.

6

AUTOMATING THE DIASPORA?

Algorithmic organisation of digital diasporas

Introduction

Automated systems and technologies actively shape and organise digital sociality. From facial recognition technologies that allow travellers to go through automated gates at the airport to recommender systems that automate the distribution of news, social media posts, online ads, which video you should view next, which route to take home, or which book you might be interested in as a result of buying a new pair of tennis shoes – many of us interact with automated media on a daily basis. Against the complex workings of these systems, digital diasporas emerge as a form of self-organising, self-realising assemblage of social relations that span the digital and material realities of everyday diasporic actors. Recommender systems that are built into social media platforms can play a significant role in creating – as well as dissolving and cutting off – social relations within these structures. It is imperative for researchers of digital diaspora to develop a robust understanding of how various layers of automation are built into the very fabric of digital sociality; digital diaspora research also needs to take into account the dynamic nature of the automated systems that enable and constrain diasporic socialities in often opaque and unexpected ways.

This chapter discusses the workings of automated systems as infrastructure for the emergence of digital diasporas and considers the methodological implications of these systems for researchers of digital diaspora. Taking the view that the algorithms that undergird digital systems are not merely technical objects (i.e., not just lines of code), but objects infused with social, cultural, and political implications, the chapter starts by reviewing the literature on algorithmic culture and discusses what it means for re-searchers of digital diaspora to foreground this complex view of technology in their research. The chapter proceeds to give a basic overview of how recommender systems work, focusing on social recommender systems and recent developments in deep learning for recommender systems. Social recommender systems are recommender systems that are specifically built for the social media domain, which play an instrumental role in helping social media users discover content, people, groups, and tags that index and

DOI: 10.4324/9781003336556-9

organise online discourse. These recommender systems are diverse in the techniques and strategies that they use in order to optimise recommendation outcomes based on a diverse range of factors – including, but not limited to community size, accuracy, diversity, freshness, fairness, time sensitivity, and overall structural robustness of a social media network. Deep learning for recommender systems is a burgeoning area of research due to increasing applications of deep neural networks trained to recognise objects in images, translate texts between languages, recognise speech, and generate images and texts from human-language prompts. Increasing interest in using deep learning to build recommender systems is attributable to its ability to process unstructured, multimedia data with minimal feature engineering efforts, as well as its potential to help solve longstanding problems in recommender systems research that can then translate into a diverse range of domains – including e-commerce, news recommendation, entertainment recommendation, and social recommendation. This chapter closes by discussing some emergent methodological issues for digital diaspora research as a result of taking seriously the automated systems and technologies that give shape to digital diasporic networks.

Algorithmic culture and automated systems as social infrastructure

Algorithmic culture as a concept was first discussed by Galloway (2006) in the context of gaming, where he considered the video game as a distinct cultural form that demands a new and unique interpretive framework. In Galloway's conceptualisation, games are algorithmic cultural objects in the sense that they are more than just fun toys; games are algorithmic machines that function through specific, codified rules of operations, and as such they are an allegory for the algorithmic structure of information culture. Beyond this game-centric articulation, several theorists have since picked up the notion of algorithmic culture to think through the increasingly central role that algorithmic systems are playing in everyday lives. Striphas (2015), for example, traced the emergence of an algorithmic culture to mean the offloading of cultural work onto computers, databases, and other types of digital technologies so that the publicness of culture is gradually abandoned, making way for the rise of an elite culture dominated by a few private entities such as Amazon, Google, and Facebook. Paul Dourish (2016), in an attempt to clarify the meaning of algorithm as an analytical object, invoked algorithmic culture to articulate the intricate embeddedness of algorithms and digital culture without equating the two. Arguing that algorithm as a concept needs to be clearly marked against closely associated concepts such as automation, code, architecture, and materialisation in order for it to maintain any useful analytical power, Dourish (2016) sought not to essentialise a sort of 'foundational truth' about algorithms, but rather to ground the analytical work that this concept mobilises within its technical and cultural milieux. Maintaining that the social sciences should seek to make interventions on the technical domains of big data practices where their technical colleagues are – that is, by forging a common language, rather than setting their own terms of reference for key terminologies such as 'algorithm' – Dourish (2016) constructed a non-social abstract space where algorithms might shape culture in very important ways but are nevertheless distinct from culture.

In this non-social abstract space, several distinctions are made in order to sharpen the analytical focus of algorithm as a concept. Algorithms are, first of all, distinct from the computer programs that may embody or implement them, in the sense that programs are both more and less than algorithms: a program might also contain non-algorithmic material, and an algorithm is more than just its specific implementation within a program (Dourish, 2016). For example, the Girvan-Newman algorithm – a community detection algorithm based on betweenness centrality (see Chapter 2) can be implemented differently in different network analysis and visualisation programs (such as Gephi) and coding libraries (such as the igraph library in R). As such, an algorithm as 'an abstract, formalised description of a computational procedure' (Dourish, 2016, p. 3) is also distinct from code, understood as human-readable expressions of program behaviour produced by programmers – or software-as-text. Algorithms are pseudo-codes in the sense that they can be operationalised in different programming languages while transcending the particularities of each of these languages and code platforms, thus maintaining their general integrity. Algorithms are not always easily and readily extracted from codes in that they can become distributed in snippets across a program, so that they are never localised and free from context. The programs that implement certain algorithms also do a lot more than just direct operationalisation of these algorithms: they read files from disks, connect to network servers, check for error conditions, respond to user prompts, record progress in log files, and so on (Dourish, 2016). Because algorithms are distributed and fragmented in a program, their position within the architecture of a computer program is oftentimes opaque: an algorithm might not be localised within even a module in a program, and since modules could be written by different programmers, running on different computers, and located within different administrative and management domains, program modules tend to be isolated from each other. Protocols that are in place to facilitate decentralised control in the writing of computer programs depend on algorithms to be implemented, but in practice, algorithms can neither be located within a stretch of code nor can they be pinpointed to a single computer or even a single organisation. As such, algorithms are also distinct from automation – argued Dourish (2016) – as regimes of digital automation encompassing systems of digital control and management through sensing, large-scale data storage, and algorithmic processing that acquires legitimacy and authority through a network of industrial, commercial, legal, and regulatory frameworks.

Largely in response to the research program that Dourish (2016) laid out, Seaver (2017) articulated a research program that understands algorithms *as* culture – so that algorithms are unstable objects enacted through the varied practices that people use to engage with them. Emphasising the analytical empowerment that comes from privileging the partial and mobile position of an outsider to algorithmic systems, Seaver (2017) made a compelling case for ethnographic enactments of algorithmic systems not as an exercise in dispelling terminological anxieties and gaining precious access to otherwise blackboxed technical systems rigidly constrained by procedural formulas, but as research tactics for a kind of ethnography that foregrounds broad patterns of meaning and practice that are constitutive of algorithms as heterogeneous and diffuse sociotechnical systems. The 'algorithms as culture' approach, argued Seaver (2017, p. 5), sees algorithms not as 'technical rocks in a cultural stream, but are rather just more water ... algorithms are cultural not because they work on things like movies or music, or because they become objects of popular concern, but because they are composed of collective human practices. Algorithms are multiple, like culture, because they *are* culture.' This formulation is particularly productive when we consider how, for example, the practice of algorithmic systems is always shifting in dynamic relation to engineers tweaking their codes to mediate between distinctive behaviours of different target groups, users trying to game the algorithm as they understand it, or people engaging with and reacting to the outcomes and outputs of algorithmic sorting and matching. This approach is also useful to think with when we consider how, as Seaver (2017, p. 7) put it, 'an algorithm's edges are enacted by the various efforts made to keep it secret' – and thus concealing the reality that algorithmic systems are populated by diverse and ambivalent characters, whose intentions and meanings cannot simply be read or guessed at. Understanding algorithms as culture gives researchers of the social sciences and humanities productive entry points with which to enter a technical discourse often shrouded in secrecy, obfuscated with jargon, and gatekept against 'nontechnical' people.

The mutual imbrication of algorithms and people means that algorithms cannot be 'revealed' so that researchers can access a stable, essential algorithmic 'core' – but rather continually unpacked so that the meaningful and performative nature of algorithms can be understood. Algorithms are temporally entrenched, argued Roberge and Seyfert (2016), since different calculations produce different outcomes, rendering algorithms always dynamic in both ontological and epistemological senses. Slack and Hristova (2021) urged social scientists to think in terms of algorithmic culture instead of algorithms, so that the connections that constitute what matters the most about the algorithm as an unstable object – its crucial integration in practices, politics, policies, economics, and everyday life – can be foregrounded. To work with algorithmic culture from the outset is to examine these elements altogether as inextricably linked, rather than isolating them and treating them separately. For Slack and Hristova (2021, p. 17), algorithms 'are best understood as complex arrangements of math, matter, and related practices, representations, experiences, affects, and effects.' The algorithmic culture paradigm seeks to never reduce sociotechnical systems to their technical components – given how they are shaped by, and actively shape, social relations.

Algorithmic culture as heuristics is useful to researchers of digital diasporas in three important ways. Firstly, it encourages digital diaspora researchers to appreciate how much of diasporic sociality is driven by algorithmic culture, and thus deserves critical attention if we are to take the question of technology seriously in our research. Secondly, algorithmic culture empowers digital diaspora researchers to engage with algorithms as an analytical object where diasporic actors and communities are; that is, as an analytical object already imbued with, and inseparable from, social life. While researchers of digital diaspora should arm themselves with as much technical knowledge as possible in order to engage with sociotechnical cultures in a robust way, they also should not feel like they need to be formally trained in engineering or computer science in order to speak with authority on technical systems that drive diasporic culture forward. Finally, the algorithmic culture paradigm encourages digital diaspora researchers to work collaboratively with colleagues in the technical domain to exchange expertise, actively contributing to the technical domain's understanding of sociotechnical systems as inherently social – and empowering technical colleagues with knowledge on how automated systems cannot be understood purely on technical terms. In the next section, I will review the emergent technical literature on recommender systems and their role in the organisation of digital resources in the spirit of the algorithmic culture approach – that is, by engaging in earnest with technical literature written often with only a specific technical audience in mind, while doing the translational work to mobilise this literature so that it becomes useful for researchers in the social sciences and humanities.

Recommender systems and the organisation of digital resources

When popular news reporting runs articles about the ever-elusive 'Facebook algorithm' (Lua, 2022) or instructs aspiring content creators on how to game the 'YouTube algorithm' (Cooper, 2021), what they are communicating is referred to in the technical literature as recommender systems. Put simply, recommender systems are 'software tools and techniques that provide suggestions for items that are most likely of interest to a particular user' (Ricci et al., 2022, p. 1). These suggestions vary depending on the particular decision-making processes at play, such as what news article to read, which song to listen to, what book to buy, or which social media post to engage with. Research on recommender systems gained traction in the mid-1990s with the rise of the web (Goldberg et al., 1992; Shardanand & Maes, 1995; Balabanović & Shoham, 1997), with the motivation of devising optimal systems to help web users select the most appropriate content given the massive amount of data available on the internet. Ricci et al. (2022) identified six different classes of recommendation approaches, including content-based, collaborative filtering, community-based, demographic, knowledge-based, and hybrid recommender systems. Content-based recommender systems recommend items by suggesting items that are similar to the ones that users have liked in the past, where similarity of items is calculated based on selected features associated with the compared items. This technique tries to match the attributes of a user profile against the attributes of an item, often in the crude form of keywords extracted from item descriptions. For example, if a user has provided a positive rating for a science fiction book, a recommender system might learn to recommend books from the same genre to this user in the future. With the collaborative filtering technique, recommender systems make recommendations based on items that other users with similar tastes have liked in the past. A prominent example of this system is Amazon's 'Customers who viewed items in your browsing history also viewed' section towards the end of a customer's homepage, or its 'Customers who bought this item also bought' recommendations. Collaborative filtering is often referred to as 'people-to-people correlation' for this reason; it is also the most popular and widely implemented technique for recommender systems (Ricci et al., 2022). In practice, item-based approaches (such as content-based techniques) are preferred in contexts where the number of users exceeds the number of items to be recommended, as they tend to produce more accurate recommendations while being more computationally efficient, requiring less frequent updates. User-based recommendations, on the other hand, tend to provide more original recommendations, which is correlated with better user experience (Ricci et al., 2022).

Community-based approaches recommend items to users based on the preferences of the users' friends. This approach is motivated by the observation that people tend to trust recommendations from their friends better than those from strangers or anonymous individuals; such an approach also relies extensively on a user's social graph, which forms the basis of the business model of social media services such as Facebook. Recommendations made based on community-based approaches rely on the ratings (a proxy for preferences) of the users' connected contacts within a social graph, such as a person's 'friends' on Facebook. This approach is particularly informative of the social recommender systems discussed in more detail below. Demographic recommender systems recommend items based on the demographic profile of a user, with the assumption that recommendations are best generated to serve demographic niches. Demographic data such as a user's language, age, gender, or ethnicity can be leveraged to direct users to different web resources. Knowledge-based systems recommend items based on specific domain knowledge tailored to a user's needs, preferences, and how much an item is useful to a user. In knowledge-based systems, a similarity function is calculated to estimate how much the user's needs (operationalised in the form of a problem description) match the recommendations (operationalised as solutions to the problem); the similarity score produced is interpreted as the utility of the recommendation for the user. Knowledge-based recommender systems often require learning components (discussed below) in order to keep up with changing user needs.

Hybrid recommender systems, as the name suggests, are systems that combine various techniques to complement the strengths and weaknesses of each recommending approach. Collaborative filtering, for example, suffers from what the technical literature refers to as the 'new-item problem' since this approach does not recommend items that have no ratings. By combining this approach with techniques of the content-based approach, which recommends items based on their descriptions, the resulting hybrid recommender system can accommodate dynamic changes in the system as new data become available. Recent advances in deep learning have also made it possible to create better performing hybrid recommender systems (Paradarami et al., 2017). The rest of this section will discuss two topics that are of central interest to researchers of digital diaspora: social recommender systems and deep learning for recommender systems.

Social recommender systems

Social recommender systems, generally speaking, are recommender systems that target the social media domain (Guy, 2022). Social recommender systems are built to recommend social media content, people to connect with, and communities (or tags and groups) for social media users to interact with. These recommender systems have important implications for digital diaspora research: diasporic actors invariably rely on these automated recommendations to find information, locate contacts, build networks, and enact diasporic performances. While social recommender systems might give the impression that they are impenetrable 'blackboxes' – given how much journalistic coverage on these systems tends to emphasise their proprietary (and thus somewhat secretive and opaque) nature, we actually know a lot about how recommender systems operate. While the specific details of TikTok's or Facebook's friend recommender algorithms, for example, are not transparent, the basics of how social recommender systems work – and thus the parameters around these systems – are entirely knowable. Furthermore, the set of content propagation algorithms on social media is relatively stable (Narayanan, 2023); these algorithms drive a large fraction of user engagement on platforms compared to search – which is when users enter search terms (keywords) to locate content. With content propagation algorithms, platforms have almost complete control over what to recommend to a user, unconstrained by what users might input as search terms. As an example, at the time of writing the most extreme case of algorithmic content propagation is TikTok's 'For You Page' (FYP) – which features content driven by what TikTok's recommender system deems most likely of interest to its users. Compare this to, for example, Facebook, Twitter, or Youtube – where content recommendation tends to be made up of a mix of algorithmic recommendations, network recommendations (content by and boosted by a person's social network), and subscription recommendations (content posted by accounts to whom users have subscribed). In an increasingly algorithmic model of social recommender system, such as TikTok's FYP, the posts a user sees are those that the algorithm predicts they are most likely to engage with – so that the *social network* component to this model is less important (Narayanan, 2023). This is an important distinction to keep in mind as researchers articulate emergent, cross-platform, multimedia digital diasporas. If the audience for each post is increasingly independently optimised based on the topic and the 'quality' of the post, making the performance of past posts by influential and/or subscribed users increasingly irrelevant, then there is less predictability and control that users can have over the reach of their content – and the kind of content they get to encounter on the platform. Could digital diasporas be said to exist on a platform such as TikTok, where the role of social networking is not very important? What are the implications of characterising digital diasporas as (not) existing on platforms that organise on the basis of algorithmic recommendations? What experimental and creative methodological approaches can be devised to meet emergent challenges with the increasingly algorithmic nature of platforms?

Social networking has played a significant role in shaping our understanding of what it means to create impactful content on social media, often expressed in terms of virality – how a piece of content spreads from person to person in a manner akin to viral contagion. As a concept, virality captures the propagation dynamic of social networks as distinct from the broadcasting dynamic of the subscription model, where content flows from one source to another unidirectionally. Goel et al. (2016) analysed a large dataset of a billion diffusion events on Twitter to formulate a formal measure called structural virality – a measure that captures variations between the two extremes of (1) content that gains its popularity through a single, large broadcast, and (2) that which grows through multiple generations with any one individual directly responsible for only a fraction of the total adoption. The structural virality of a post is the number of degrees of separation, on average, between users in the corresponding tree. The deeper the tree, with more branches, the greater the structural virality. Even though structural virality gives us a good understanding of the variety and degrees of structural virality, it is very hard to predict the virality of social media content based on the information available when the content is published, such as the content of a social media post and information about the creator of the post. The virality of a post is also subject to platform demotion – often referred to as 'shadowbanning' by social media users – a practice in which content deemed problematic is shown to fewer users, but not removed from the platform (Narayanan, 2023). Much of the anxieties around the non-transparency of social recommender systems can be said to come from this idea of shadowbanning, because lower count of engagement and reach is not automatically or directly attributable to platform demotion (given how there are always many factors that go into the engagement optimisation of each post). Diasporic actors who rely on the reach and engagement of the content they produce in order to build up communities and make profit from providing services for their communities often already have very elaborate and dynamic heuristics about how platform recommendation algorithms work for or against them; these heuristics, in turn, contribute to the feedback loop that has been built into social recommender systems, which rely on user behaviour data for engagement optimisation.

Before machine learning became standard practice for social recommender systems, Facebook used to implement an algorithm called EdgeRank to construct a user's feed. EdgeRank, which was replaced by machine learning algorithms designed to optimise a metric called 'meaningful social interactions' (MSIs), worked along two main dimensions: affinity score and item type weight. Affinity score represents Facebook's prediction of how much a user wants to see updates from a given poster and is manually programmed to take into account metrics such as whether the user recently interacted with the poster (Narayanan, 2023). Item type weight is also manually set to reflect what type of media is expected to be more engaging, such as photo or video (livestreamed or uploaded) over text. With MSIs, the predicted probability that a user will have a specific type of interaction (such as 'reacting' or commenting) with a given item is calculated with machine learning, where manual weights for item types like photos or videos are no longer necessary since preference for content type can be automatically learned from data on a per-user level. That MSIs still use affinity scores (a social network metric) instead of fully implementing a machine learning recommender system that optimises for immediate engagement, Narayanan (2023) observed, seems to be a deliberate attempt on Facebook's part to manually program a preference for friends-and-family content into the platform, so that while short-term engagement might be compromised, long-term satisfaction could be preserved. The weights given to platform metrics such as 'reacts', reshares, or comments are constantly tweaked in response to issues that emerge as a result of public and scholarly scrutiny over network outcomes. MSIs that give significant weight to comments, reshares, and direct messages ended up giving priority to content that provoked outrage and stoked division, and Facebook's documented attempt to counter this by reducing the weights given to reshare while still privileging comments seems to also have backfired, as content creators carried on creating divisive content in ways that would elicit a lot of comments (Narayanan, 2023).

As we think through how digital diasporas are formed and what sort of diasporic content gets prioritised and becomes the materials with which diasporic socialities are enacted, researchers should actively think with platform affordances – such as recommender systems – as well as the wider socio-political context in which platforms operate. It is also empowering to encourage research participants to think through these systems and how they amplify certain modes of sociality over others during the research process; once we acknowledge that automated systems are inherently social, it becomes intuitive to open up these systems to discussion and exploration by the diverse publics whose lives are entangled in complex ways within these systems. Research that blackboxes these automated systems, or treats them as inevitable, risks entrenching the opacity and unfairness that they have been criticised for – while only making partial and biased headway towards understanding what digital diasporas might be made of. Digital diaspora researchers should also commit to staying abreast of emergent advances in recommender systems as well as dynamic platform implementations of these advances so that the ongoing project of assembling diasporic socialities can be kept up to date. In the section below, we will review some recent advances in deep learning for recommender systems – and discuss the wider implications of deep learning as a paradigm of automation for diasporic socialities.

Deep learning for recommender systems

Deep learning is a subset of machine learning techniques and approaches that utilise simulated neural networks with multiple layers to learn and represent complex patterns and relationships in data (Mitchell, 2019). Simulated neural networks are modelled after the structure and function of the human brain, and are comprised of a large number of interconnected processing nodes (neurons) that work together to perform complex tasks such as image recognition, natural language processing, and predictive analytics. Neural networks often have an input layer, several hidden layers where an activation function is computed, and an output layer. In this multilayered architecture, each neuron receives input from other neurons and uses that input to compute a weighted sum, which is then used to generate an output after passing through hidden layers. This output becomes the input to other neurons in the network, and the process repeats until a final output is generated. This design mimics the way that biological neurons signal to one another, and is inspired by insights from neuroscience about the human brain.

Major neural networks include multi-player perceptrons (MLPs), convolutional neural networks (CNNs), recurrent neural networks (RNNs), autoencoders, generative adversarial networks (GANs), and graph neural networks (GNNs) (Zhang et al., 2019). These networks have been variously applied to improve and solve recurrent problems in recommender systems. GANs, for example, are a type of deep learning model that consists of two neural networks: a generator and a discriminator. GANs have been used in recommender systems to generate personalised recommendations. An example of GAN-based recommender system is the Generative Adversarial User Model (GAUM) model, which generates personalised recommendations by learning the user-item preference distributions using an adversarial training procedure in which the generator network generates new user-item pairs while the discriminator network evaluates the quality of the generated pairs (Chen, 2019). GNNs have also been increasingly applied in recommender systems due to their ability to handle structured data and explore high-order information. GNNs can alleviate the problem of data sparsity in recommender systems with semi-supervised signals, such as assisting in modelling the target behaviours with other behaviours in multi-behaviour recommendations (Gao et al., 2022).

A topic of central interest to digital diaspora researchers is how diasporic actors stay informed about and connected with 'home' and across the diasporas. Online news services increasingly deploy news recommender systems to help readers explore news resources by personalising news content based on user behaviour – often categorised into click and dwell behaviours – and incorporating both long-term and short-term user representations to capture users' evolving interests and preferences over time (An et al., 2019; Wu et al., 2019). Reinforcement learning-based approaches use a reward-based system to train the recommendation model, where the model is rewarded for recommending articles that the user interacts with and penalised for recommending articles that the user ignores or dislikes (Afsar et al., 2022). Reinforcement learning incorporates information about a user's historical behaviour – such as articles read and articles shared, as well as contextual information, such as time of day, location, and device – in the design of its 'state space', which represents the recommender environment. The 'action space' represents the set of all possible actions that the agent can take in a given state in response to the state it is in, which feeds into the 'reward space', which represents the set of all possible rewards that the agent can receive as a result of its actions in a given state. The reward function maps each state-action pair to a numerical reward value, which provides feedback to the agent on the quality of its actions (Afsar et al., 2022). How these dynamic systems work to inform and shape diasporic views on topics spanning political and cultural identities as well as senses of belonging, are extremely understudied; not much is known, also, on how diasporic engagement with these systems on the ground is shifting diasporic socialities in a more granular and measured level – that is, beyond broad notions of 'staying informed' and 'staying connected'.

Recommender systems and digital diasporas: some emergent methodological issues

This chapter has given an overview of the workings of automated systems as infrastructure for the emergence of digital diasporas, as well as outlining the ways in which recommender systems act as important cultural intermediaries that organise what people encounter when they engage with digital services. The chapter has also argued for the importance of engaging with automated media as sociotechnical systems – that is, as complex systems that cannot be reduced to either their social or technical dimensions. The digital has always been at once social and technical; by overlooking the ways in which the technical is intensely social in its practice – and how the social is entangled with the technical through layers of automation that have been built into the architecture of digital diasporas – researchers risk complicity in reproducing misguided and unequal narratives around the legitimacy and importance of research into automated systems.

Four emergent methodological issues can be articulated here by way of concluding this chapter (and this book more generally). The first issue deals with the theoretical and methodological implications of studying digital diasporas as embedded within, and intimately constitutive of, algorithmic cultures – rather than as a social product of technical arrangements and affordances. Foregrounding this important framework means that digital diaspora research should not render technological systems static and hidden in the background, while only choosing to highlight what can be directly observed through participant observation, interviews, or even ethnographic studies that focus solely on diasporic interactions *on* digital platforms. Important dimensions of algorithmic cultures include hardware (mobile devices, computers, cameras, internet infrastructures, computing infrastructures) as well as software (web browsers, apps, platforms, algorithms), both understood in their various contexts. The most innovative studies and research programs to emerge within digital diaspora studies will have to contend with the complexities of disentangling and reassembling these more-than-human assemblages as a way forward. Thinking about digital diasporas as algorithmic cultures requires a commitment to think about the sociotechnical as a unit, irreducible to its constituent parts – and its heterogenous components inseparable from each other.

The second issue deals with the need to rethink digital diaspora studies as an interdisciplinary and multidisciplinary field of inquiry. While digital diasporas have gathered significant interest from researchers working across the humanities and social sciences and sparked important cross-collaborations among these fields, there is an urgent need for digital diaspora researchers to engage with, and bring in researchers from, the more technical disciplines such as computer science and engineering. Colleagues in the technological sciences as well as technicians who directly design the architecture of digital diasporas often work under conditions where there is little institutional and professional incentive to engage with the perspectives and insights of the humanities and social sciences. Digital diaspora researchers not only need to learn as much technical knowledge about automated systems as possible – so that they can interrogate, complicate, and problematise the assumptions built into this body of knowledge – but also engage in collaborative ways with their colleagues working in the technical sciences. Designing methodologies that incorporate techniques and approaches from computer sciences can enrich digital diaspora studies in very exciting ways; empowering our colleagues in the technological sciences with the knowledge of how sociotechnical systems cannot be abstracted away from everyday practices can also help open up multidisciplinary conversations around ethical, responsible, and inclusive technology design.

The third issue concerns the methodological challenges of studying automated systems that are increasingly algorithmic in their logic as they move away from the social network model. While it remains to be seen whether the algorithmic model of recommender systems will become the predominant paradigm of organising social media content, the challenges of designing the appropriate methods to study them are a pressing concern of the present. How are digital diasporas being enacted in the presence of these algorithmic systems? What path dependency might exist between platforms that enable and constrain diasporic socialities in unexpected ways? How do recommendation algorithms amplify or repress diasporic voices, and what might be the implications of these processes? Digital diaspora research will have to be increasingly interdisciplinary to answer these important questions; researchers will need to engage in earnest with the latest developments in recommender system research and ongoing debates in the political economy of technologies in order to embed their accounts of diasporic socialities within the sociotechnical contexts of their unfolding.

Finally, even though it is beyond the scope of this book to discuss creative research methods, it is becoming increasingly important for digital diaspora researchers to engage with diasporic research participants in ways that not only foreground their experiences, but also empower them to understand, critique, and actively shape the automated systems that drive digital diasporas forward. There is a strong tradition of arts-based research which draws on the artistic process in all forms, such as written, visual, spoken, and performance, which has enabled researchers to explore experiences of a topic in collaboration with research participants. These methods are particularly generative in their capacity to enable researchers and their participants to articulate existing imaginaries about technological systems – so that new imaginaries can take shape and be productively built out of various creative prompts and expressions. In actively dispelling myths about automated systems as part of their research design, researchers not only can create positive impact on the communities they are studying, but also inch ever closer to the heart of the phenomenon they wish to study. By embracing and bringing together methods from different research traditions, digital diaspora researchers can advance the field in exciting ways, forge new research programs, and generate deeper understandings about the ways in which technology and society intersect and shape each other. By cultivating collaboration and mutual respect with research participants, researchers can build stronger relationships and foster a sense of shared ownership over the research process. This, in turn, can lead to more nuanced and insightful findings, as participants can offer unique perspectives and experiences that might have been overlooked otherwise. The innovative potential for digital diaspora research is vast, and by taking an open and collaborative approach, researchers can unlock new insights, build bridges between disciplines and communities, and create a more vibrant and inclusive field.

References

Afsar, M. M., Crump, T., & Far, B. (2022). Reinforcement learning based recommender systems: A survey. *ACM Computing Surveys*, *55*(7), 1–38. 10.1145/3543846.

An, M., Wu, F., Wu, C., Zhang, K., Liu, Z., & Xie, X. (2019). Neural news recommendation with long-and short-term user representations. In *Proceedings of the 57th annual meeting of the association for computational linguistics* (pp. 336–345).

Balabanović, M., & Shoham, Y. (1997). Fab: content-based, collaborative recommendation. *Communications of the ACM, 40*(3), 66–72.

Burke, R. (2007). Hybrid web recommender systems. In P Brusilovsky, A Kobsa, & W Nejdl (Eds.), *The adaptive web: methods and strategies of web personalization* (pp. 377–408). Springer.

Chen, X., Li, S., Li, H., Jiang, S., Qi, Y., & Song, L. (2019). Generative adversarial user model for reinforcement learning based recommendation system. In *International conference on machine learning* (pp. 1052–1061).

Cooper, P. (2021, April 18). How the youtube algorithm works in 2023: The complete guide. *Hootsuite*. https://blog.hootsuite.com/how-the-youtube-algorithm-works/.

Dourish, P. (2016). Algorithms and their others: Algorithmic culture in context. *Big Data & Society, 3*(2), 2053951716665128. 10.1177/2053951716665128.

Galloway, A. R. (2006). *Gaming: Essays on algorithmic culture* (Vol. 18). University of Minnesota Press.

Gao, C., Wang, X., He, X., & Li, Y. (2022). Graph neural networks for recommender system. In *Proceedings of the fifteenth ACM international conference on web search and data mining* (pp. 1623–1625).

Goel, S., Anderson, A., Hofman, J., & Watts, D. J. (2016). The structural virality of online diffusion. *Management Science, 62*(1), 180–196. 10.1287/mnsc.2015.2158.

Goldberg, D., Nichols, D., Oki, B. M., & Terry, D. (1992). Using collaborative filtering to weave an information tapestry. *Communications of the ACM, 35*(12), 61–70. 10.1145/138859. 138867.

Guy, I. (2022). Social recommender systems. In F Ricci, L Rokach, & B Shapira (Eds.), *Recommender systems handbook* (Third Edition). Springer, NY, USA.

Lua, A. (2022, May 24). Decoding the Facebook Algorithm in 2023: A fully up-to-date list of FB Algorithm changes and best practices. *Buffer*. https://buffer.com/library/facebook-news-feed-algorithm/.

Mitchell, M. (2019). *Artificial intelligence: A guide for thinking humans*. Penguin, UK.

Narayanan, A. (2023, March 9). Understanding social media recommendation algorithms. *Knight First Amendment Institute at Columbia University*. https://knightcolumbia.org/content/understanding-social-media-recommendation-algorithms.

Paradarami, T. K., Bastian, N. D., & Wightman, J. L. (2017). A hybrid recommender system using artificial neural networks. *Expert Systems with Applications, 83*, 300–313. 10.1016/j.eswa.2017.04.046.

Ricci, F., Rokach, L., & Shapira, B. (2022). Recommender systems: Techniques, applications, and challenges. In F Ricci, L Rokach, & B Shapira (Eds.), *Recommender systems handbook* (3rd ed., pp. 1–35). Springer.

Roberge, J., & Seyfert, R. (2016). What are algorithmic cultures? In R Seyfert, & J Roberge (Eds.), *Algorithmic cultures: essays on meaning, performance and new technologies* (pp. 13–37). Routledge.

Seaver, N. (2017). Algorithms as culture: Some tactics for the ethnography of algorithmic systems. *Big Data & Society*, 4(2), 2053951717738104. 10.1177/2053951717738104.

Shardanand, U., & Maes, P. (1995). Social information filtering: Algorithms for automating "word of mouth". In *Proceedings of the SIGCHI conference on Human factors in computing systems* (pp. 210–217).

Slack, J. D., & Hristova, S. (2021). Why we need the concept of algorithmic culture. In S Hristova, S Hong, & J D Slack (Eds.), *Algorithmic culture: How big data and artificial intelligence are transforming everyday life* (pp. 15–34). Lexington Books.

Striphas, T. (2015). Algorithmic culture. *European Journal of Cultural Studies*, 18(4–5), 395–412. 10.1177/1367549415577392.

Wu, C., Wu, F., An, M., Qi, T., Huang, J., Huang, Y., & Xie, X. (2019). Neural news recommendation with heterogeneous user behavior. In *Proceedings of the 2019 Conference on Empirical Methods in Natural Language Processing and the 9th International Joint Conference on Natural Language Processing (EMNLP-IJCNLP)* (pp. 4874–4883).

Zhang, S., Yao, L., Sun, A., & Tay, Y. (2019). Deep learning based recommender system: A survey and new perspectives. *ACM Computing Surveys (CSUR)*, 52(1), 1–38. 10.1145/3285029.

APPENDICES

Appendix 1

TABLE 3.6 Topics generated and their top twenty terms

Topic	Share % M (SD)	Top twenty terms (translated)
Topic 1: TM in context of formal national healthcare	3.33 (0.67)	Limited time, Curcumin, Seniority, Ranking, Two-week period, Bcl-2 protein family, Tràng phục linh (Colitis medication), Namo amitabud, Bảo an khang (Name of insurance scheme), Cutleaf groundcherry, Sickness, Tiger bone glue, Gathering, Half portion, Short course, Corn, Dental care, Russia, Kidney stone, Geoduck
Topic 2: General principles of TM	3.84 (1.02)	Local dialect, Quality, Ease, Human organs, Equilibrium, Blood veins, Medical effects, Fruits, Human anatomy, Egg, Cashew, Selection, Alternative, Kidney, Vinegar, Good quality, Pepper, Immersion, Softness, Food dish
Topic 3: Knowledge sharing as advertising for direct sales	2.67 (0.57)	Symptom, Dragon's tongue leaf, Phlegm, Service, Human back, Tuber fleeceflower, Southern medicine, Tomorrow, Once a day, Circumstances, Plant processing, Plant scouting, Inbox, Knowledge, Medicine, Parasite, Recipe, Soda, Lymphatic system, Frequency
Topic 4: TM as religious philanthropy	3.31 (0.18)	Seafood, Frequency, Healing, Buddhist monk, Fruit, Phlegm, Basic, Consequences, Philippines, Yao people, Species, Hua Tuo, [Redacted username], Favour, Variety, Tree trunk, Plucking, [Redacted username], Sharpness

(*Continued*)

TABLE 3.6 (Continued)

Topic	Share % M (SD)	Top twenty terms (translated)
Topic 5: General childcare advice	2.21 (0.46)	Blanket, Hotness, False daisy, Homegrown, Fibraurea leaf, Table, Standard, Correction, Determination, Ground substance, Average, Seeing, Thread, Malignant hyperthermia, Chickenpox, Addiction, Childhood, Vân Hồ (a district in Sơn La, Vietnam), [Redacted username], Feces
Topic 6: TM as lifestyle	4.39 (1.04)	Blowing, Multitude, Avoidance, Shower, Tickling, Discovery, Leanness, Gypsum, Moisture, Sleep, Territory, Udumbarra flower, Backside, [Redacted usersame], [Redacted username], Atherosclerosis*, Vinegar*, Prevention, Fine meal, Miracle
Topic 7: Dietary benefits of traditional plants and call for direct sales	4.26 (1.42)	Sabah snake grass, Digestion problems, [Redacted username], Contrast, Ray, Year, Condensed, Tonsillitis, Pneumonia, Instructions, [Redacted username], Tomorrow, Soup, Red blood cell, Effectiveness, Zalo (a Vietnamese messaging application), This, Body, Gift, [Redacted username]
Topic 8: TM as dietary supplements	3.21 (0.44)	Vietnamese ginseng, Brand name, An Tôn (a former village in Vĩnh Phúc district, Thanh Hoá province), Alleviation, Negativity, Hihi, Vitamin B12, Body temperature, Sympathy, Small dots, Familiarity, Chinese mesona, Falling, Western, Fairy, Cashew, Cornea, Half portion, Intermediate, Men
Topic 9: TM as narratives	4.02 (1.32)	Medicality, Hour of the pig, Desire, 30 minutes, Publishing request, Once upon a time, Brothers and sisters, Death, Heat, Stop, Flanovoid*, Pencil cactus, Steaming, Cannabis, Seeds, Paper-thin, Medicine, Ingredients, Country, The Americas
Topic 10: TM in family health	3.75 (1.02)	Rice wine, Lightness, Eternity, Itchiness, [Redacted username], [Redacted username], Bastard children, Tropics, Warmth, Calmness, Sand, Cashew, Sweetness, Computer, Beauty, Menstrual cramps, Tumour, Household equipments, Substance
Topic 11: Ethnic variations in TM	1.09 (0.43)	Peeling, Gold apple, Career, Cardboard, Socks, Body part, Support, Satisfaction, Lâm Thao (a Northern province in Vietnam, also name of a fertiliser brand), [Redacted username], Solasodine*, Peony, Thuốc Nam Phong (rheumatism medication), Promotion, Noel*, Infection, Doctor Triệu Thị Thanh (a Yao doctor), Clots*, Magnesium, Zinc

(Continued)

TABLE 3.6 (Continued)

Topic	Share % M (SD)	Top twenty terms (translated)
Topic 12: Managing alcohol addiction	2.14 (0.98)	Wood, [Redacted username], Body, Sickness, Pearl, Retrogade ejaculation, Baldness, Darkness, Trick, Movement, Bad luck, Thuỷ Ngàn (alcohol addiction medication), Bad temper, Vitamin K, Confession, Infection, Folliculitis, Craziness, Below*, Hemorrhoids
Topic 13: Shaming uses of TM	1.30 (0.40)	Beta-Sitosterol*, Shrimp paste, Speed, Pomade, Technology, Space*, Sterculiaceae*, 6 years, Monoglucoside*, Rarity, Precipitation, Supression, News, Mood, Bookstore, Country bumpkin, Reasoning, Morning sickness, Occasionality, Belief
Topic 14: Managing mental health among overseas Vietnamese	2.23 (0.34)	America, Wormwood, Rowatinex*, Hatred, Juice, Contact, Heaviness, Thankfulness, Isothiocyanate, Insomnia, Anxiety, Kohlrabi, Rhythm, Sunglasses, Translation*, Participation, Bruises, Borrowing, Vacuum, Sickness
Topic 15: Managing smoking addiction with TM	2.36 (0.26)	Antonin Seipin* (supposedly Russian doctor), Sleep, Vontaren*, [Redacted username], Gấc seeds, Bird, Walking, Dog blood, Hunger, 70 years old, Rheumatism, Water, Spine, Chicken, Picture, Gift, Food, Thankfulness, Heaviness, Intern
Topic 16: Cardiovascular health and overseas Vietnamese	2.26 (0.42)	Dick*, [Redacted username], Kkkkkkk, Untimely diagnosis, Shred, Penis, Garden, Combination, Announcement, Contradiction, Urging, Vitality, Hole, Hematology, Acceleration, Withdrawal, Visa*, Blood veins, Good reputation, Cikan (heart health supplement)
Topic 17: Anti-infective plants and pain management	2.00 (0.48)	La Gi (small town in Bình Thuận, Southern Central Vietnam), Common Purslane, Admin, Surplus, Usefulness, Salonpas, Indigenousness, Natural antibiotics, Summer, Autumn, Aches, Flower, Catching wind, Exercise, Trần Hưng Đạo (name of a street), Mink*, Bloodline, 100k, Boil
Topic 18: Cautionary tale against misuse of TM	1.57 (1.07)	Body, Women, Buying, Asking, Extra, Wine, Strength, Usage, Cancer, Diagnosis, Death, Sugar, Men, Friend, Steambath, Weekly, Turmeric, Tea, Derris
Topic 19: Men's sexual & reproductive health	3.42 (1.44)	[Redacted username], Dry-zone mahogany bark, Expenses, Cure, Scabies, Effectiveness, [Redacted username], Researcher, Materials, Archipelago, Regret, Prevention, Amber, Gossypol*, Indian goosegrass, Testicles, Limonene*, Hygiene, Schefflera heptaphylla, Pedology

(Continued)

TABLE 3.6 (Continued)

Topic	Share % M (SD)	Top twenty terms (translated)
Topic 20: Food as medicine	3.84 (0.24)	Food with 'cold' properties, Fermented tofu, Money, Duck meat, Area, Cash flow, Contradiction, Sickness, Interaction, Diaper, Method, Water, Bear gallbladers, Springing, Medical practice, Pig, Mother, Reference, Honey, Hybridity
Topic 21: Agrarian lifestyle as healthy lifestyle	2.28 (0.28)	Mushroom, PDR* (Physicians's Desk Reference), Happiness, Moment, [Redacted username], Tongue, 50 years, Buffalo, Quảng Ninh (a Northern province in Vietnam), Boil, Salt, Flies, Fullness, Applying ground leaves over wound, Shampoo, Class of plants, Fruit tree, Tea, Eating habits, Từ Sơn (a village in Bắc Ninh, a Northern province)
Topic 22: Buddhist lifestyle as healthy lifestyle	1.94 (0.12)	Suffocation, Digestion, Vegetarianism, Piper lolot leaf, Proof, Cần Thơ (a province in Southern Vietnam), Namo amitabud, Tree stump, Black bean, Studying overseas, Brain, Today, [Redacted username], The truth, Boat*, Taking medication, Storytelling, Adrenal gland, Happiness and Prosperity, Body organ
Topic 23: Discussions on the origins and originality of Vietnamese medicine	3.14 (1.73)	[Redacted username], [Redacted username], Patent, Clutching, Usage, Southern territories, [Redacted username], Better quality, Ethnic minorities, Lips, Determination, Admin approval, Destiny, Tuệ Tĩnh Đường Liên Hoa (a TM dojo in Huế, Central Vietnam), Trịnh Hoài Đức (18th century historical figure), [Redacted username], Coach bus, Malt, Pounding
Topic 24: Processes of preparing plants as medicine	1.62 (0.29)	House yard, Đống Đa (a district in Hà Nội), 6 months, Chromatophore, Purity, Daytime, Itchiness, Sleep, Nutrition, Rice wine, Oxygen, Chaff-flower, Gastrict acid, Bowl, Citizen, Fermenting and Drying, Purpose, Afternoon, Ribs, [Redacted username]
Topic 25: Emergency childcare advice	1.37 (0.43)	Tools and equipments, Baby colic, Examination, Squeezing, Amount, Inner body, Selection, Sharp pain, Binystar (baby colic medication), Liking, Roseola, Couch, Lung, Dollar, [Redacted username], [Redacted username], Alkaloid*, Statistics, Ceylon hill gooseberry, Mothers
Topic 26: Negotiating sales of medicinal plants	2.35 (0.44)	Hedyotis*, Willow bark, [Redacted username], Parasite, Personal taste, Family recipe, Blue, Middle, White, Negativity, Experience, Lingao County, Savings, Colitis, Eel, Storm, Feet, Virginity, Post office, Quality

(Continued)

TABLE 3.6 (Continued)

Topic	Share % M (SD)	Top twenty terms (translated)
Topic 27: Negotiating shipping methods and sales of medicinal plants	2.13 (0.72)	[Redacted username], Shade trees, Bank, 350k, Carboxymethyl cellulose, Peanut, Miracle, Flagellate, Hà Nội, Incomprehensibility, Index finger, Stemona tuberosa, Concurrence, Long Biên (a district in Hà Nội), Dear friend, Billards, Hoàng Công (coach bus brand), Square, Birds, Wrapping cloth
Topic 28: Gastroenterology health	2.62 (0.51)	Recipe, Sugar beets, Investment, [Redacted username], Business, Gastralgia, Eating*, Falling off, Country, Family recipe, Discussion, Flatulence, Macau, [Redacted username], Urination, Guava, Stinkvine, Effect, [Redacted username], Money
Topic 29: Mental health and longevity	2.00 (0.62)	Reason, 100-year-old, [Redacted username], Greetings, Finickiness, Reasoning, Well-wishing, Anolyte, Acute pain, [Redacted username], Oneself, Sending, The truth, Quảng Ninh (a Northern province in Vietnam), Bait, Costliness, Heaviness, Jokes, Spa
Topic 30: Aging and Loneliness	1.42 (0.23)	Belonging, Myself, Sesame, Animal bones, Chi Thống Hoàn (osteoarthritis medication), Cadmium, Diaphoretic, Drawing, Blood, Dr. Lê Minh, Viettel (a telecommunication company), Envelope, Shampoo, Ramnoza, Kangaroo, Traditional medicine street, Lipstick, Prescription, Earth, Cat
Topic 31: Old age and health	1.29 (0.39)	Produce, Virus, Grind, Ringworm, Hypertension, Quality, Pleiku (a city in the Central highlands), Food dish, Unit, Coughing, Stew, Everybody, Water pipe, My aunt, Sickness, Manhood, Items*, Pharmacist, [Redacted username], Bad temper
Topic 32: Manging the common cold with TM	2.65 (0.28)	Sky, Capture, Labour, Weight, Panax pseudoginseng, Plant family, Samurai (energy drink brand), Conclusion, [Redacted username], Acyclovir, Buying, Cupping therapy, [Redacted username], The academy, Bravery, Treasure, Majority, Master
Topic 33: Nutrition and cardiovascular health	2.05 (0.19)	Battleground, Soil, Voice, Bran, Tip*, Nun, Quality, Ease, Human organs, Equilibrium, Blood veins, Medical effects, Fruits, Human anatomy, Egg, Cashew, Selection, Alternative, Kidney, Vinegar
Topic 34: Healthy lifestyle and religious narratives	2.02 (0.44)	Winter melon detox juice, Zona, Job's tears seeds, Strawflower tea, Freshness, Observation, Dharmapala, Afternoon, Namo

(Continued)

TABLE 3.6 (Continued)

Topic	Share % M (SD)	Top twenty terms (translated)
		amitabud, Inauguration, The moon*, Fruit, Instance, Beauty, Hibiscus, Medicinality, Multitude, Mint, Adults, Share*
Topic 35: Men's health and cardiovascular health	2.03 (0.60)	Actuality, Chia seeds, Spora Lygodii, Mass communication, Positivity, Magnolia bark, Bracelet, [Redacted username], Flatulence, Sudden, [Redacted username], High endogenous testosterone, Sharing, Total occlusion, Infection, Certainty, Guava, 10 kg, Water, Basic
Topic 36: Women's beauty and sexual health	1.02 (0.16)	[Redacted username], Chlorophyll, Helping, Body, Know-how, Soul, Droppers*, Counsellor, Indochinese serrow, [Redacted username], Senior Colonel, Obstetric, Tetronic acid*, Euphorbia ambovombensis, Forrest, North winds, Flabby, Secret code, Ficus*
Topic 37: Narratives of medical families and family recipes	4.01 (1.57)	Efficacy, [Redacted username], False ginseng, Family recipe, Household registration book, Goods, Caterpillar fungus powder, Infertility, [Redacted username], Tour*, Retaining, The passing of spring, [Redacted username], Flower, Chickrassy, Water caltrops, Listed price, Wind, Filtered water, Aches, Rice
Topic 38: Insomnia and discussion of burnout	1.37 (0.22)	Disinfection, Wooden floor, Week, Sock*, Burning pain, 'Bread and butter', Year of the Dog, [Redacted username], Virus, Inadequate sleep, Gum, [Redacted username], Jaundice, Root cause, International*, [Redacted username], Caligan*, 330 mg, Willow tree, Jelly
Topic 39: Diet and women's health	3.34 (1.33)	Snake, Forgotten recipes, Multitude, Winged bean pods, Women, Coconut shell, Once upon a time, Waistline, Weight loss, Ignoring, Rice paddy herb, Hypertension, Vestibular disorders, Shellfish, Toxaemia, Đà Nẵng (a city in Central Vietnam), Salt, [Redacted username], Virgin fish sauce
Topic 40: Pregnancy advice	2.94 (0.55)	Gypsum, Choking on a fishbone, Liver, Cornea, Helicteres hirsuta Lour, Long Ju, Flatulence, Aiming, Từ Dũ (an obstetric hospital in Ho Chi Minh City), Afterhour shirts, City, Night, Manufacturer, [Redacted username], [Redacted username], Pomade, [Redacted username], Kilogram, Common cold during pregnancy, Legitimacy

(Continued)

TABLE 3.6 (Continued)

Topic	Share % M (SD)	Top twenty terms (translated)
Topic 41: Health and beauty tips	2.10 (0.13)	Indian goosegrass, Flower stigma, Someday, Past recipes, Sarsi, Rice paddy herbs, Skills, Member*, Gifting, Blood cockle, Five fruits, Water caltrop, Brother and sisters, Cafeine, Salt, Professional, Delivery, [Redacted username], [Redacted username], Homegrown
Topic 42: Cautionary tales against abuse of indigenous tobacco	1.41 (0.20)	Coronary artery disease, Experience, Thuốc rê (traditional rustic tobacco), [Redacted username], Eggplant, U Minh (commune in Cà Mau, Southernmost province in Vietnam), [Redacted username], Almond, Wisdom teeth, Anticipation, Miracle, Monk fruit, Asthma, Phú Thọ (province in Northern Vietnam), Vestibular disorders, Step, Gấc seeds, Positivity, Origins, Sinusitis
Topic 43: Fantastic tales about the religious and historical origins of Vietnamese medicine	2.73 (0.76)	Miniscule, Today, Baton, Annoyance, Northern Central, Ākāśagarbha (a Buddhist Bodhisattva), Regret, Đà river (Northern Vietnam), The way, [Redacted username], [Redacted username], Himalayas, Needles, Joy, Trưng sisters (ancient history women warriors), Snakehead, Time, Panadol, Accidentality, [Redacted username]
Topic 44: Traditional alternatives to biomedicine and overseas Vietnamese	1.22 (0.19)	Friend, Germany, [Redacted username], Miniscule, Dandruff, Military, Apple, Operation, Nature, Anti-inflammatory, Muscovy duck, Testicles, Timeliness, Heat, Pregnancy, Baby, mmol/L*, Oysters, [Redacted username], Multitude
Topic 45: Constipation and hot/cold binary	4.45 (0.78)	Sickness, Orchid, Loneliness, Orphan, Tightness, Extract, Sharing, Alcohol*, Goose, List, Sapodilla, Hygiene, Remainder, [Redacted username], [Redacted username], [Redacted username], Blue*, Eyedrops, Gulan*, Cooling agent, Purple heart plant
Topic 46: Northern medicine and haemostasis	4.02 (1.09)	Reduction, 50 cm, Bleeding, Red beans, Pebbles, Phú Xuyên (a district in Hà Nội), Tip*, Pangolins, Pueraria thomsonii flower extract, Concurrency, Activity, Parasite, Thorns, Cover, Fish, Fungi, Health, Steaming, 1 month, Weighing scale
Topic 47: Nutritions and women's health	3.35 (1.72)	Withdrawal, Overreaction, Name, Mutuality, Aches, Mentality, Jar, Early, Long process, Infection, Sinusitis, Women, Pharmacy, Pouring, Sisters, Soaking, Time, Symptoms, Ming aralia, Crinum latifolium

(*Continued*)

TABLE 3.6 (Continued)

Topic	Share % M *(SD)*	Top twenty terms (translated)
Topic 48: Pain management with TM	1.78 (0.77)	Raising, Gratitude, Daisy, X-ray, Spinal disc herniation, Hospital, Phoenix eyes, Rambutan, Myself, Dry blood, Gauze, Advice, Cabbage, [Redacted name], Criticism, Quantifying, Superior grade, Poaching, Truthfulness, Sliding
Topic 49: Otolaryngology and TM	1.12 (0.07)	Photograph, Gum, Nose, Care, White, [Redacted username], Cabbage, Bone, Lemon, Buttox, Conclusion, Minority, Life, Bottle cap, Sea, Vitamin B, Children, Papaya, Belching, Hot temper

Note: Terms are translated into English where appropriate. Proper nouns (brand names, location names) are kept in Vietnamese, accompanied by explanations in brackets. Common names of plants are preferred over their scientific names, although not all plants have common names in English. Usernames are redacted to ensure anonymity. Terms that were originally written in a language other than Vietnamese are marked with '*'.

Appendix 2 Summary statistics for the interpretation of a topic

Topic 30 – Aging and loneliness

Top words	
$\lambda = 1$	$\lambda = 0.6$
Sesame	Belonging
Animal bones	Myself
Prescription	Sesame
Earth	Animal bones
Belonging	Chỉ Thống Hoàn (osteoarthritis medication)
Dr. Lê Minh	Cadmium
Myself	Diaphoretic
Cat	Drawing
Envelope	Blood
Viettel	Dr. Lê Minh
Shampoo	Viettel (a telecommunication company)
Chỉ Thống Hoàn (osteoarthritis medication)	Envelope
Diaphoretic	Shampoo
Kangaroo	Ramnoza
Lipstick	Kangaroo
Cadmium	Traditional medicine street
Drawing	Lipstick
Blood	Prescription
Viettel (a telecommunication company)	Earth
Ramnoza	Cat

Rank-1 metric: rank 37 out of 70
Coherence metric: rank 23 out of 70

Note: This statistic presentation is modelled after Maier et al (2018). The figure depicts a divided table and an inter-topic distance map, where the specific topic in summary is coloured red. The table maps out the top-words according to two different relevance values ($\lambda = 1$ and $\lambda = .6$). Below the table, the ranks of the Rank-1 and the coherence metrics are given.

Inter-topic distance map (via multidimensional scaling using 'LDAvis' package)

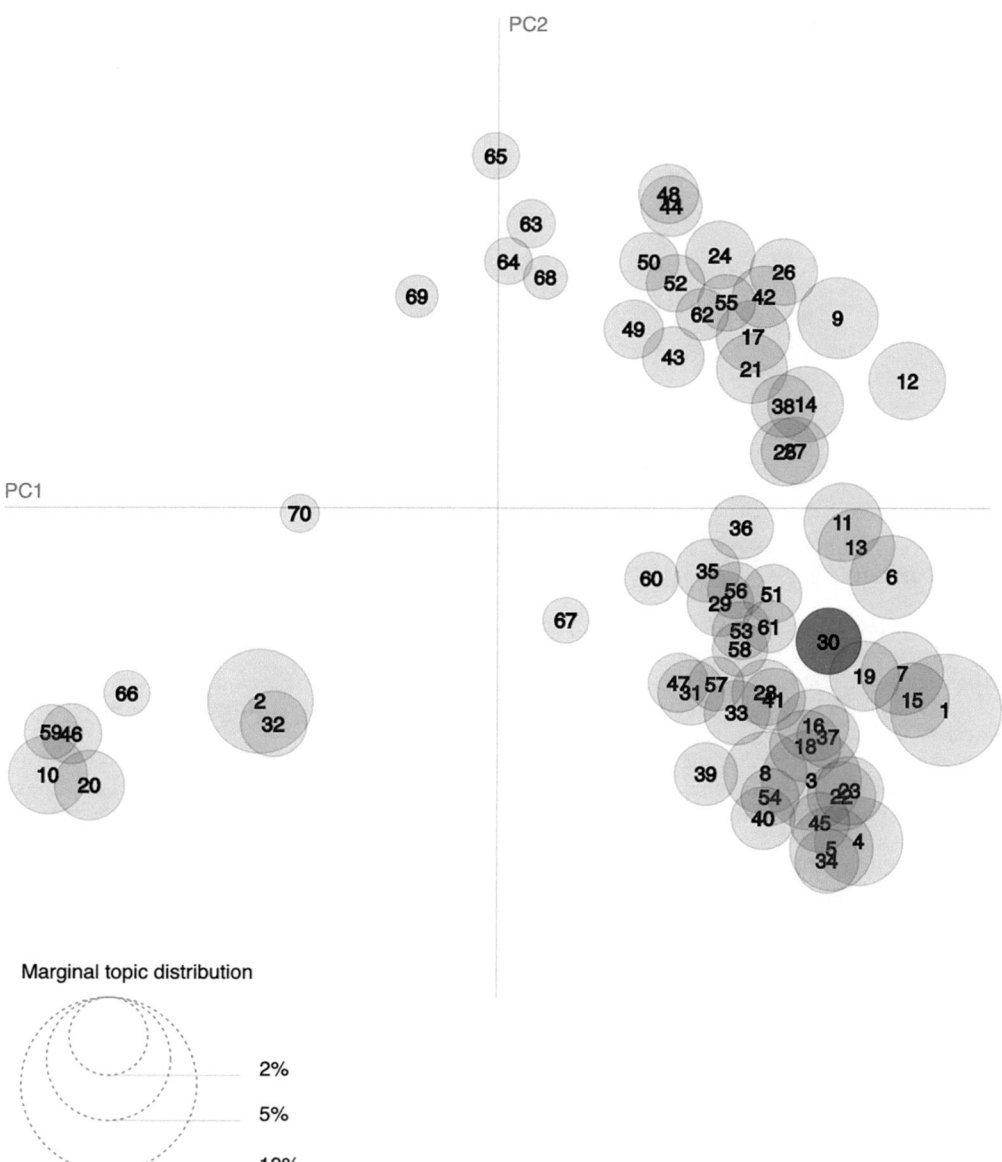

Marginal topic distribution

2%

5%

10%

INDEX

Page numbers followed by "n" indicate notes; page numbers in **bold** indicate tables; page numbers in *italics* indicate figures